T0192828

Textile Science and Clothing Technology

Series editor

Subramanian Senthilkannan Muthu, SGS Hong Kong Limited, Hong Kong, Hong Kong

Subramanian Senthilkannan Muthu
Editor

Textiles and Clothing Sustainability

Recycled and Upcycled Textiles and Fashion

 Springer

Editor
Subramanian Senthilkannan Muthu
Environmental Services Manager-Asia
SGS Hong Kong Limited
Hong Kong
Hong Kong

ISSN 2197-9863   ISSN 2197-9871   (electronic)
Textile Science and Clothing Technology
ISBN 978-981-10-9541-2   ISBN 978-981-10-2146-6   (eBook)
DOI 10.1007/978-981-10-2146-6

Printed on acid-free paper

This Springer imprint is published by Springer Nature
The registered company is Springer Science+Business Media Singapore Pte Ltd.

Contents

Fashion Renovation via Upcycling

Thilak Vadicherla, D. Saravanan, M. Muthu Ram and K. Suganya

Abstract Sustainability in fashion is a big challenge, and textile waste can be used as the raw material for value-added products. A framework has been developed in the present chapter for the creation of upcycled products from garment waste. Ten case studies have been depicted pictorially. Indian and global current scenario on upcycled garments along with the challenges in upcycling is also discussed.

Keywords Sustainability · Fashion · Garment waste · Framework · Upcycling

1 Introduction

Of late, the term "sustainability" has become a buzzword across the globe. Sustainable is defined by the Oxford dictionary as "able to be maintained at a certain rate or level" or "conserving an ecological balance by avoiding depletion of natural resources".[1] A first of its kind definition for the term sustainable

---

[1]http://www.oxforddictionaries.com/definition/english/sustainable.

T. Vadicherla (✉) · D. Saravanan · M. Muthu Ram · K. Suganya
Department of Textile and Fashion Technology, Bannari Amman Institute of Technology, Sathyamangalam, Erode, Tamilnadu, India
e-mail: thilak.vadicherla@gmail.com

D. Saravanan
e-mail: dhapathe2001@rediffmail.com

M. Muthu Ram
e-mail: muthurammuthaiah@gmail.com

K. Suganya
e-mail: suganyakanagaraj93@gmail.com

© Springer Science+Business Media Singapore 2017
S.S. Muthu (ed.), *Textiles and Clothing Sustainability*,
Textile Science and Clothing Technology,
DOI 10.1007/978-981-10-2146-6_1

development was suggested by Brundtland (formerly known as World Commission on Environment and Development), as "the development that meets the needs of the present without compromising the ability of future generations to meet their own needs" (United Nations 1987). The United Nations Environment Programme (UNEP) predicts that an enormous amount of stress is going to be witnessed on the depleting scarce natural resources, considering the rate of consumption that would become three times higher by the year 2050 with the present rate of consumption being taken into account (Annual Report 2011). The simplest understanding of existence of sustainability is that rates of natural resources generation and consumption should match. Classic 3R concept that comprises reuse, reduce and recycle is regarded as one of the widely acknowledged solutions in the sustainable development. 4R concept for the benefit of sustainable development has been evolving with the addition of the element named "rebuy" to the classic 3R concept. 4R concept highlights the significance of rebuying the products manufactured from the reused or recycled or reclaimed materials.

2 Fast Fashion

Fashion can be seen as an expression of one self which is widely accepted by a group of people over the time and has been characterized by several marketing factors such as shorter life cycle, low predictability, high impulse purchase and high volatility of market demand (Fernie and Azuma 2004). Till 1980s, fashion industry's popularity and success was focused on low-cost mass production of standardized styles with the exception being haute couture (Brooks 1979). Since 1999, fashion shows and catwalks became a public phenomenon which led to demystification of the fashion process.[2]

Wide spread use of technology combined with the surge of internet and digitization has enabled rapid spreading of information/trends that results in consumer's ability to have a lot of options (Economist 2005). Consumers demand for more and more styles in shorter span of time has inspired retailers like Zara, H&M, Mango, New Look, and Top Shop to adopt designs rapidly and quickly introduce interpretations of the runway designs to the stores in a minimum of three to five weeks (Hoffman 2007). Fast fashion is a term used to describe cheap and affordable clothes which are the result of catwalk designs moving into stores in the fastest possible way in order to respond to the latest trends.[3] Fast fashion retailing is leading consumers towards an increased rate of purchasing and the trend to keep clothing for an ever shorter time with the resulting rise in clothing disposal. Fashion is about being trendy, up to date and latest, and sustainability is about long lasting, durable, low impact and eco-friendly.

---

[2]http://www.sydneylovesfashion.com/2008/12/fast-fashion-is-trend.html.
[3]http://www.macmillandictionary.com/dictionary/british/fast-fashion.

3 Textile Waste and Recycling

3.1 Classification of Textile Waste

Textile wastes materials may be broadly classified into three categories[4] viz, (i) pre-consumer textile wastes (PrCTW), (ii) post-industrial textile wastes (PITW) and (iii) post-consumer textile wastes (PtCTW). Pre-consumer textile wastes (PrCTW) are those wastes which never make it to the consumers and which come directly from the original manufacturers. Examples include ginning wastes, opening wastes, carding wastes, comber noils, combed waste yarns, roving wastes, ring spinning waste fibres, ring spun waste yarns, open-end spinning waste fibres, open-end spinning yarn wastes, knitting waste yarns, weaving waste yarns, fabric cutting wastes, fabric wet processing wastes and apparel manufacturing wastes. Post-industrial textile wastes (PITW) are generated during the manufacturing process of upstream products. These are mainly from the virgin fibre producers, tire cord manufacturers, polymerization plants and other plastic products. Post-consumer textile wastes (PtCTW) are the wastes that come from the consumer, and these are generally the clothes that are ready for disposal or landfill. They are recovered from the consumer supply chain. Favourite examples of the PtCTW include recycling of the accessories and beverage bottles to make recycled polyester.

3.2 Recycling Technologies

Recycling technologies (Scheirs 1998; Wang 2006) are divided into primary, secondary, tertiary and quaternary approaches based on the raw materials used and the products produced at the end of the process (Table 1).

Primary recycling involves recycling a product into its original form example being industrial scraps; secondary recycling involves mechanical (melt) processing of post-consumer plastic product into a new product that has a lower level of physical, mechanical and/or chemical properties. Tertiary approach involves processes such as pyrolysis and hydrolysis, which convert the plastic wastes into basic

Table 1 Recycling approaches

Approaches	Raw material for recycling
Primary approach	Industrial scraps
Secondary approach	Mechanical processing of post-consumer products
Tertiary approach	Pyrolysis/hydrolysis of polymeric wastes to get monomers or fuels
Quaternary approach	Burning the fibrous solid wastes and utilizing the heat generated

---

[4]http://www.recycling.about.com.

chemicals or monomers or fuels. Quaternary recycling refers to burning the fibrous solid waste and utilizing the heat generated.

4 Textile Upcycling

4.1 History of Upcycling

The US EPA estimates that textile waste occupies nearly 5 % of all landfill space and textile recycle industry recycles approximately 15 % of all PCTW and leaves 85 % in landfills.[5] Upcycling and downcycling can be considered as examples of recycling. The term "upcycling" was coined by Reiner Pilz of Pilz GmbH in 1994 significantly who explained the concept of adding value to the old or used products, which is quite contrary to the popular concept of recycling that reduces the value of the products (Aa). The first book published on upcycling was authored by Gunter Pauli in German language in the year 1998 and was adapted by Johannes F. Hartkemeyer and then Director of the Volkshochschule in Osnabruck. William McDonough and Michael Braungart's book "Cradle to Cradle: Remaking the Way We Make Things", published in the year 2002, caught the attention of public and cemented its place in public usage.

Typically, upcycling creates something new and better from the old or used or disposed items. Process of upcycling requires a blend of factors like environmental awareness, creativity, innovation and hard work and results in a unique sustainable and handmade product. Upcycling aims at the development of products truly sustainable, affordable, innovative and creative. For example, downcycling produces cleaning rags from worn T-shirts, whereas upcycling recreates the shirts into a value-added product like unique handmade braided rug. Literature reveals that few fashion industries[6] are already into the foray of recreating value-added products using upcycling from used materials or pre-loved materials or dead stock or left over materials and/or employing socially sustainable worker/work practices.

4.2 Global Scenario

4.2.1 Kallio

Kallio was founded by Karina Kallio, which is a kid's wear brand, whose clothes are known to be fun and functional with a green heart (see Footnote 6). Kallio recrafts men's dress shirts into stylish, modern classics for kids aged infant to

---

[5]http://www.weardonaterecycle.org/.

[6]http://kallionyc.com.

8 years old. The garments developed by Kallio are recreated from used vintage clothing, which are popular for their unique colour, pattern and design. Kallio garments are developed using hand cut and recrafted them into kid's wear at their location in New York, USA.

4.2.2 Sword & Plough

Sword & Plough is considered as a socially conscious brand that recycles military surplus fabric into stylish purses and bags.[7] It works with American manufactures which employ veterans and donate 10 % of profit back to veteran organizations. The notable feature of this brand is its commitment to empowering veterans with dignified employment and green environment. Since the bags are made from military fabric, they are known to last long and are "rugged and refined".

4.2.3 Reformation

Reformation is a Los Angles-based Fashion Company founded by Yael Aflalo in 2009 and aims at creating sexy and sleek styles for women.[8]. Raw materials at Reformation include new sustainable textiles, repurposed vintage clothing and rescued dead stock fabric from fashion houses that over-ordered. Their collection caters to mainly multiple women who are between 5'6 and 5'10, and a petite collection is designed for ladies 5'4 and under. Unique feature of Reformation is "all under one roof" aspect, which includes design, manufacture, photograph and selling from the same warehouse and factory. Reformation is committed to sustainability and uses recycled paper, energy efficient lighting, non-toxic cleaning supplies and sustainable business practices.

4.2.4 Looptworks

Looptworks was founded in 2009 by Scott Hamlin and Gary Peck that aims at sustainable enterprise.[9] The working principle of Looptworks is "rescuing high-quality, unused material and converting it into something beautiful and useful for your everyday life in the name of limited edition, hand-numbered goods. Looptworks' products include preexisting neoprene to create sleeves and cases for tech products such as tablets and laptops, backpacks and women's and men's apparel. Looptworks is committed to sustainability by becoming one of the partners for Southwest Airlines' Luv Seat: Repurpose with Purpose project, which upcycles

---

[7]www.swordandplough.com
[8]www.thereformation.com
[9]www.looptworks.com.

80,000 leather seat covers leftover from old airline interiors into new products such as special edition bags.

4.2.5 Seamly.co

Seamly.co was founded by Kristin Glenn, with an intention to produce apparel responsibly, with thought and soul and care.[10] This brand lays emphasis on the process and uses surplus fabrics (excess from other factories and designers, knitted in the USA, or sustainably and responsibly made overseas.) and entire garmenting process happens domestically.

4.2.6 Reclaimed

Reclaimed is an eco-friendly and socially responsible USA social enterprise that creates one-of-a-kind upcycled dresses from vintage or vintage-inspired clothing.[11] It was founded in 2014 and creates high-quality dresses for stylish, socially conscious fashionistas and supports nonprofits that educate and empower women and girls in the developing world.

4.3   Indian Scenario

4.3.1   Trmtab

Trmtab was founded by Mansi and Cassandra, which utilizes leather scraps from factories around the world to create limited edition, and refined leather goods for tech devices.[12] The name Trmtab is inspired by a term coined by systems theorist Buckminster Fuller which means the little pressure on the ship rudder can change the direction of a ship. Similarly, an individual may be the starting point of a big impact. Trmtab launched the factory's first upcycling initiative in partnership with Prachi Leathers, in Kanpur, India, and wants to be the small change.

4.3.2   BlueMadeGreen

Ratna Prabha Rajkumar, who runs an upcycling boutique, imaginatively called BlueMadeGreen, from her residence in Bangaluru.[13] The name was derived from

---

[10]www.seamly.co.

[11]www.etsy.com/shop/ReclaimedFashion.

[12]www.trmtab.com.

[13]http://her.yourstory.com/prabhas-bluemadegreen-0116.

the denims, which are basically in blue shade being converted to value-added products terming them as green. Some of the products done by her include shopping bags, backpacks, laptop bags, handbags, sling bags, denim skirts, jackets, frocks, aprons, organisers, cushion covers and bedspreads, to name a few.

4.4 Framework for Upcycling

A framework has been proposed for the utilization of garment waste. Various factors such as type of waste, amount of value addition to be given to the product, impact of value addition, cost of the product, societal impact, environmental impact and consumer behaviour have been considered in the proposed framework. Proposed framework is shown below.

1. Selection of garment or garment waste.
2. Identification of defects in the garment.
3. Creation of design.
4. Evaluation of garment with the design.
5. Construction of the garment.
6. Significance of the product developed.

The above framework is utilized for the purpose of creating ten upcycled garments from the garments or garments waste and is discussed.

5 Case Studies on Textile Upcycling

5.1 Case Study 1—Conversion of Ladies T-shirt into Kid's Nightwear

Step 1: **Selection of Garment**

- Select a damaged garment from the pre-consumer waste of apparel industry for upcycling.
- Selected garment: ladies T-shirt (Fig. 1).

Step 2: **Identification of Defects**

- Identify the defects/damages in the garments.
- Defect: operation missing (neck rib missing) (Fig. 2).

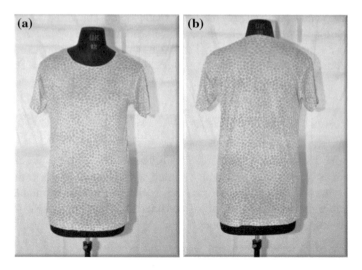

Fig. 1 Damaged/defective garment—ladies T-shirt. **a** Front view and **b** back view

Fig. 2 Garment defect (neck rib missing)

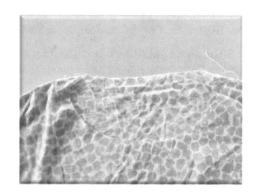

Step 3: **Creation of Design**

- Generate ideas to upcycle the pre-consumer waste (ladies T-shirt) from various researches.
- Design those ideas into the form of illustration/sketches.
- Analyse whether the design can able to replace/alter the defects in final product.
- Sketch the original product and final output (before and after).
- Design kid's night wear with ruffles at bottom and pleated straps (Fig. 3).

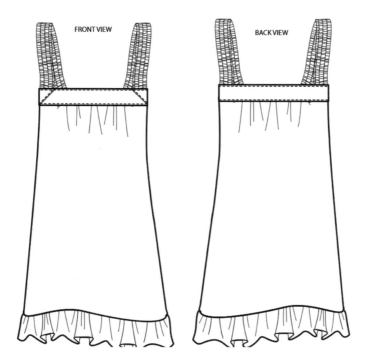

Fig. 3 Design creation of kid's night wear with ruffles at bottom and pleated straps

Table 2 Kid's size chart

Age	Height	Chest	Waist	Hip
4 years	38–42″	22″	21″	23″

Step 4: **Evaluation of the Garment with Design**

- Select a garment from the pre-consumer waste which must be evaluated by matching the following requirements.
- The garment must able to match the sizing measurement of the final output.
- Evaluate the garment dimensions to make alternation and construction (ease allowance and seam allowances, etc., are to be considered).
- Ladies T-shirt size: large
- Kid's night wear size chart is shown in Table 2.

Step 5: **Construction of the Garment**

1. Cut out the side seam along with sleeves and separate the front and back panels.
2. Cut out the defects (neck part) from both front and back panels.
3. Shape the neck and armhole according to the size and design (Fig. 4).

Fig. 4 Defective garment
after marking as per new
design

Fig. 5 Defective garment
after cutting as per new design

4. Finish the raw edges of the neck and armhole using binding (Fig. 5).
5. Make ruffles using the waste back neck portion which was removed at initial stage (Fig. 6).
6. Attach the ruffles with the front and back of the bodice portion and sew the side seam (Fig. 7).
7. Make pleated straps using the waste pieces from the sleeve portion (Fig. 8).
8. Attach the straps in the bodice portion.
9. Do the necessary surface ornamentation.
10. Kid's night wear is prepared from the ladies T-shirt (Fig. 9).

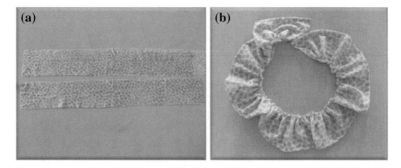

Fig. 6 Preparation of ruffles. **a** Cut portions and **b** ruffle

Fig. 7 Attachment of ruffle with the front, back and side seam

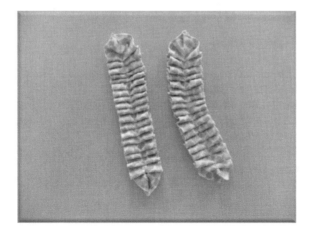

Fig. 8 Preparation of pleated straps

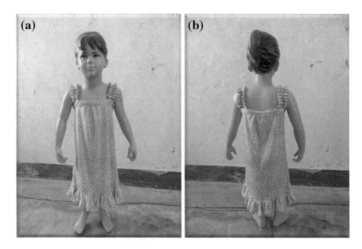

Fig. 9 Upcycled garment. **a** Front view and **b** back view

5.2   Case Study 2—Conversion of Ladies T-shirt and Tank Top into Kid's Frock

Step 1: Selection of Garment

Selected garments: (1) ladies T-shirt and (2) ladies tank top (Fig. 10).

Step 2:   **Identification of Defects**

Defects in ladies T-shirt:

1. Printing mistake.
2. Open seam.

Defect in tank top: slubs (Fig. 11).

Step 3:   **Creation of Design**

- Generate ideas to upcycle the pre-consumer waste (ladies T-shirt) from various researches.
- Design those ideas into the form of illustration/sketches.
- Analyse whether the design can able to replace/alter the defects in the final product.
- Sketch original product and final output (Before and After).
- Design kid's frock.

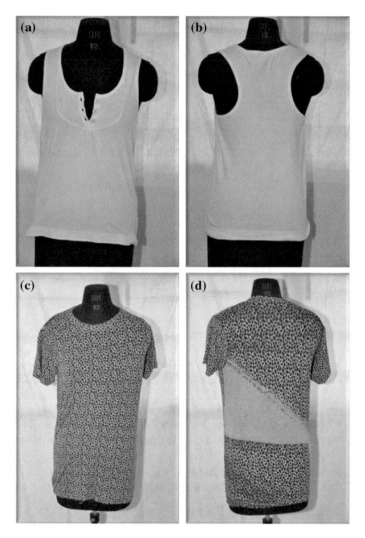

Fig. 10 Damaged/defective garments. Ladies T-shirt **a** front view and **b** back view. Ladies tank top **c** front view and **d** back view

Step 4: **Evaluation of the Garment with Design**

- Select a garment from the pre-consumer waste which must be evaluated by matching the following requirements.
- The garment must able to match the sizing measurement of the final output.

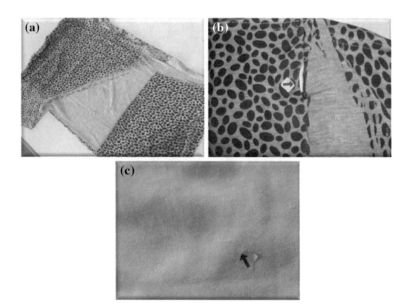

Fig. 11 Garment defects. **a** Printing mistake, **b** open seam and **c** slub

Table 3 Size chart

Age	Height	Chest	Waist	Hip
4 years	38–42″	22″	21″	23″

- Analyse the garment dimensions that are sufficient to make alternation and construction (Keep in mind about the ease allowance, seam allowance).
- Ladies T-shirt size: large.
- Kid's frock size: size chart (Table 3).

Step 5:   **Construction of the Garment**

1. Remove the damaged areas in both the garments. Remove the slubs area in the tank top and remove the miss printed area from the T-shirt (Fig. 12).
2. Merge the garment according to the design.
3. Shape the garments according to the design and remove the unwanted pieces (Fig. 13).
4. Now sew the raw edges areas of the top portion by using tank top wasted raw edges parts to finish it (Fig. 14).
5. Produce gathers in the bottom portion of frock (tank top) (Fig. 15).
6. Attach the top (T-shirt) and bottom portion (tank top).
7. The kid's frock is converted using ladies T-shirt and tank top (Fig. 16).

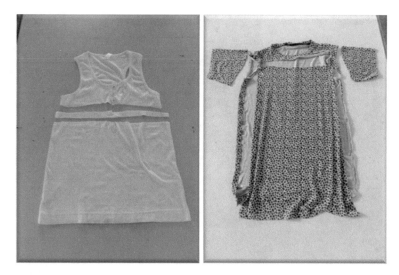

Fig. 12 Defective garment after cutting

Fig. 13 Defective garment
after cutting and arranging as
per new design

Fig. 14 Sewing of raw edges
of top

Fig. 15 Preparation of
gathers in the bottom portion
of frock

5.3   Case Study 3—Conversion of T-shirt into Kid's a-Line Skirt

Step 1:   **Selection of Garment**

Selected garment: T-shirt (Fig. 17).

Step 2:   **Identification of Defects**

- Defects: 1. Uncovered hems. 2. Misprints (Fig. 18).

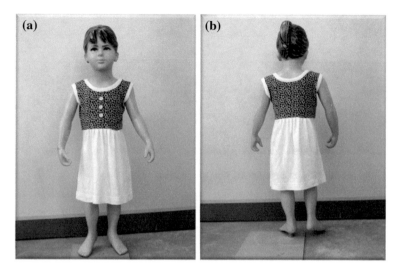

Fig. 16 Upcycled garment. **a** Front view and **b** back view

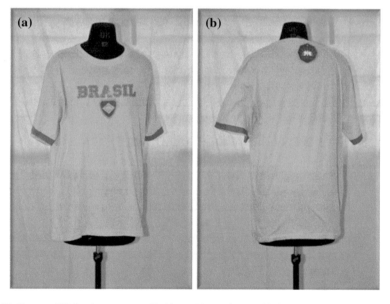

Fig. 17 Damaged/defective garments T-shirt. **a** Front view and **b** back view

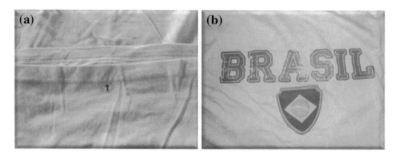

Fig. 18 Garment defects. **a** Uncovered hems and **b** misprints

Table 4 Size chart

Age	Height	Chest	Waist	Hip
4 years	38–42″	22″	21″	23″

Step 3: **Creation of Design**

- Generate ideas to upcycle the pre-consumer waste (ladies T-shirt) from various researches.
- Design those ideas into the form of illustration/sketches.
- Analyse whether the design can able to replace/alter the defects in final product.
- Sketch original product and final output (before and after).
- Design kid's A-line skirt.

Step 4: **Evaluation of the Garment with Design**

- Select a garment from the pre-consumer waste which must be evaluated by matching the following requirements.
- The garment must able to match the sizing measurement of the final output.
- Analyse the garment dimensions are sufficient to make alternation and construction (Keep in mind about the ease allowance, seam allowance).
- Ladies T-shirt size: large.
- Kid's frock size (Table 4).

Step 5: **Construction of the Garment**

1. Cut a piece of fabric from the T-shirt for making skirt. Below picture shows how to cut the garment for making A-Line Skirt (Fig. 19).
2. Cut a piece from the back shoulder side to use for waist band for skirt and cut the uncovered damaged hem portion at bottom (Fig. 20).

Fig. 19 Defective garment
after cutting as per new design

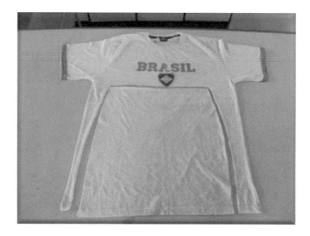

Fig. 20 Preparation of
waistband

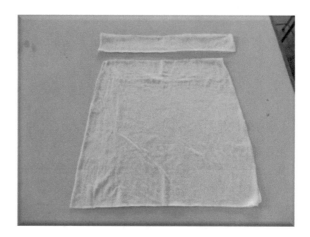

3. Produce gather and attach the waistband with the skirt portion (Fig. 21).
4. Hem the bottom of the skirt and sew the side seam.
5. Add the necessary embellishment work. We added a flower design using the sleeve bottom portion and made a pocket on back with stone works.
6. Hence the kid's A-line skirt is converted from the damaged T-shirt (Figs. 22 and 23).

Fig. 21 Preparation of gathers and attachment of waistband with skirt

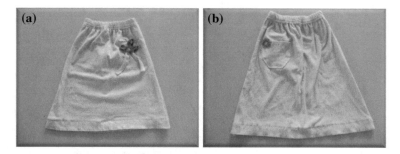

Fig. 22 Kid's A-line skirt. **a** Front view and **b** back view

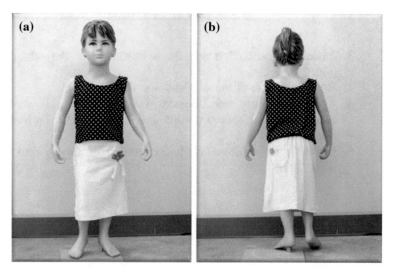

Fig. 23 Kid's A-line skirt with the top. **a** Front view and **b** back view

5.4 Case Study 4—Conversion of Nighty and Frock into Kid's Frock

Step 1: **Selection of Garment**

Selected garments: (1) nighty and (2) frock (Fig. 24).

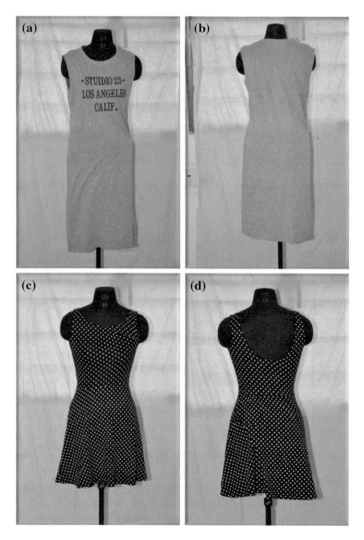

Fig. 24 Damaged/defective garments. Nighty **a** front view and **b** back view. Frock **c** front view and **d** back view

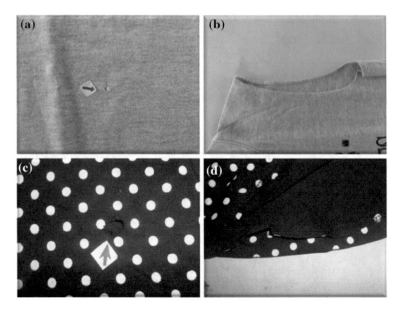

Fig. 25 Defects in nighty. **a** Holes and **b** incomplete/missing operation. Defects in frock **c** holes and **d** uncovered hems

Step 2: **Identification of Defects**

Defects in nighty: (1) holes and (2) incomplete/missing operation.
Defects in frock: (1) holes and (2) uncovered hems (Fig. 25).

Step 3: **Creation of Design**

- Generate ideas to upcycle the pre-consumer waste (ladies T-shirt) from various researches.
- Design those ideas into the form of illustration/sketches.
- Analyse whether the design can able to replace/alter the defects in final product.
- Sketch original product and final output (before and after).
- Design kid's frock.

Step 4: **Evaluation of the Garment with Design**

- Select a garment from the pre-consumer waste which must be evaluated by matching the following requirements.
- The garment must able to match the sizing measurement of the final output.
- Analyse whether the garment dimensions are sufficient to make alternation and construction (Keep in mind about the ease allowance, seam allowance).

Table 5 Size chart

Age	Height	Chest	Waist	Hip
4 years	38–42″	22″	21″	23″

Fig. 26 Cutting of required fabric from nighty

- Ladies T-shirt size: large.
- Kid's frock size chart is shown in Table 5.

Step 5: **Construction of the Garment**

1. Remove the damaged parts in the nighty and cut the required amount of fabric from the garment for making kid's frock (Fig. 26).
2. Cut piece of square shape fabric from the remaining chest portion for making yoke in kid's frock (Fig. 27).
3. Shape the garment according to the design (Fig. 28).
4. Attach the yoke part with the bodice panel and hem the bottom and sew the side seam (Fig. 29).
5. Remove the damages from defected frock and cut it into pieces of squares for making ruffles (Fig. 30).
6. By using the above pieces of defect frock, make the ruffles as well as make the straps for the kid's frock (Fig. 31).
7. Attach the straps in the kid's frock (Fig. 32).
8. Attach the ruffles at the bottom and also make a flower design at the yoke and knee portion (Figs. 33 and 34).

Fig. 27 Preparation of yoke

Fig. 28 Shaping of garment according to design

Fig. 29 Attachment of yoke
part with bodice and hem the
bottom and sew the side seam

5.5 Case Study 5—Conversion of T-shirt into Kid's Frock

Step 1: **Selection of Garment**

- Selected garment: T-shirt (Fig. 35).

Step 2: **Identification of Defects**

- Defect: hole (Fig. 36).

Step 3: **Creation of Design**

- Generate ideas to upcycle the pre-consumer waste (ladies T-shirt) from various researches.
- Design those ideas into the form of illustration/sketches.
- Analyse whether the design can able to replace/alter the defects in final product.
- Sketch original product and final output (before and after).
- Design kid's frock.

Step 4: **Evaluation of the Garment with Design**

- Select a garment from the pre-consumer waste which must be evaluated by matching the following requirements.
- The garment must able to match the sizing measurement of the final output.

Fig. 30 **a** Removal of
damaged parts from frock and
b preparation of ruffles

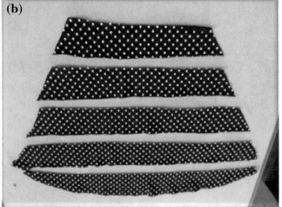

- Analyse whether the garment dimensions are sufficient to make alternation and construction (Keep in mind about the ease allowance, seam allowance).
- Ladies T-shirt size: large.
- Kid's night wear size (Table 6).

Step 5: **Construction of the Garment**

1. Lay the garment and make marks for cutting.
2. Cut the neck portion along with placket (Fig. 37).
3. Cut the garment into panels according to the design (Fig. 38).
4. Attach the neck portion along with the chest and sleeve panels (Fig. 39).
5. Now make gathers at the bottom portion.
6. Attach the bottom gathers along with the bodice panel.
7. Trim the unnecessary threads.
8. Hence the kid's frock is converted from T-shirt (Figs. 40 and 41).

Fig. 31 a Preparation of
straps and **b** ruffles

5.6   Case Study 6—Conversion of T-shirt into Baby Frock

Step 1:   **Selection of Garment**

Selected garment: T-shirt (Fig. 42).

Step 2:   **Identification of Defects**

- Identify the defects/damages in the garments.

Defects: (1) holes and (2) stain

Fig. 32 Attach straps to kid's frock

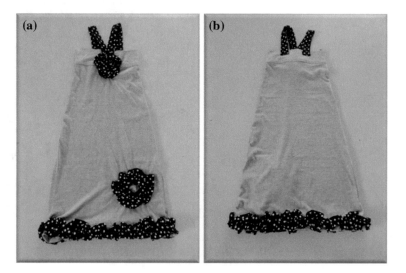

Fig. 33 Surface embellishment to the kid's frock. **a** Front view and **b** back view

Step 3: **Creation of Design**

- Generate ideas to upcycle the pre-consumer waste (ladies T-shirt) from various researches.
- Design those ideas into the form of illustration/sketches.
- Analyse whether the design can able to replace/alter the defects in final product.

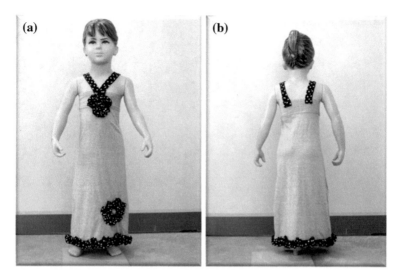

Fig. 34 Upcycled garment. **a** Front view and **b** back view

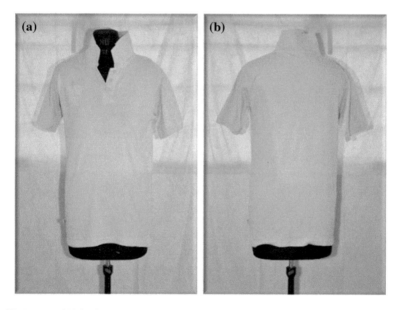

Fig. 35 Damaged/defective garment—T-shirt. **a** Front view and **b** back view

Fig. 36 Garment defects—
hole

Table 6 Size chart

Age	Height	Chest	Waist	Hip
4 years	38–42″	22″	21″	23″

Fig. 37 Defective garment
after cutting the neck portion

Fig. 38 Defective garment
after cutting as per new design

Fig. 39 Attach the neck portion along with the chest and sleeve panels

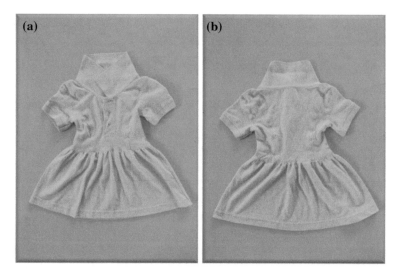

Fig. 40 Upcycled garment. **a** Front view and **b** back view

- Sketch original product and final output (before and after).
- Design baby frock.

Step 4: **Evaluation of the Garment with Design**

- Select a garment from the pre-consumer waste which must be evaluated by matching the following requirements.
- The garment must able to match the sizing measurement of the final output.
- Analyse whether the garment dimensions are sufficient to make alternation and construction (Keep in mind about the ease allowance, seam allowance).

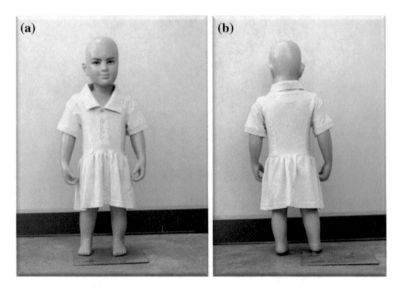

Fig. 41 Upcycled garment on the dummy. **a** Front view and **b** back view

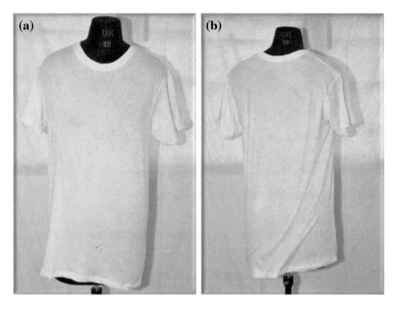

Fig. 42 Damaged/defective garment—T-shirt. **a** Front view and **b** back view

Table 7 Size chart

Age	Height	Chest	Waist	Hip
4 years	38–42″	22″	21″	23″

Fig. 43 Defective garment
after removing unwanted
pieces

Fig. 44 Preparation of fabric
for making strap for frock

- Ladies T-shirt size: large.
- Kid's night wear size (Table 7).

Step 5: **Construction of the Garment**

1. Cut the unwanted pieces from the T-shirt (Remove the chest and sleeve parts) (Fig. 43).
2. Take a piece of fabric from the cut out waste fabric for making strap for frock (Fig. 44).

Fig. 45 Making of gathers at
the chest area

Fig. 46 Hem the bottom and
sew the side seam

3. Remove the damaged parts from the T-shirt.
4. Produce gathers at the chest area on both sides of the frock, according to the design (Fig. 45).
5. Hem the bottom and sew the side seam (Fig. 46).
6. Attach the straps with the frock.
7. Hence the T-shirt is converted into baby frock (Fig. 47).
8. Add the necessary embellishment work (Fig. 48).

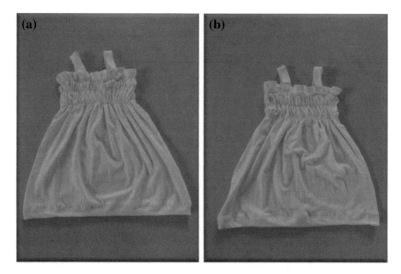

Fig. 47 Upcycled garment (baby frock). **a** Front view and **b** back view

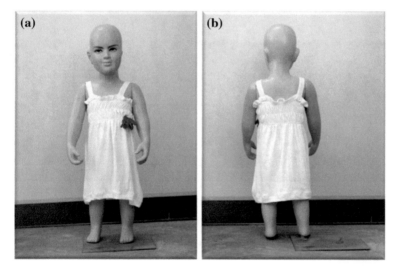

Fig. 48 Upcycled garment (baby frock) on the dummy—**a** front view and **b** back view

Fig. 49 Damaged/defective garment—T-shirt. **a** Front view and **b** back view

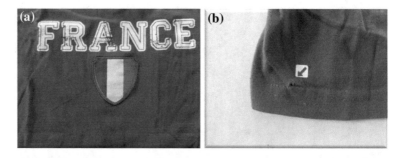

Fig. 50 Garment defects. **a** Misprint and **b** open seam

5.7 Case Study 7—Conversion of T-shirt into Kid's Fancy Frock

Step 1: **Selection of Garment**

  • Selected garment: T-shirt (Fig. 49).

Step 2: **Identification of Defects**

  • Identify the defects/damages in the garments.
  • Defects: (1) misprint and (2) open seam (Fig. 50).

Step 3: **Creation of Design**

- Generate ideas to upcycle the pre-consumer waste (ladies T-shirt) from various researches.
- Design those ideas into the form of illustration/sketches.
- Analyse whether the design can able to replace/alter the defects in final product.
- Sketch original product and final output (before and after).
- Design kid's fancy frock with one side sleeve.

Step 4: **Evaluation of the Garment with Design**

- Select a garment from the pre-consumer waste which must be evaluated by matching the following requirements.
- The garment must able to match the sizing measurement of the final output.
- Analyse whether the garment dimensions are sufficient to make alternation and construction (Keep in mind about the ease allowance, seam allowance).
- Ladies T-shirt size: large.
- Kid's night wear size (Table 8).

Step 5: **Construction of the Garment**

1. Lay the garment and mark the garment for cutting.
2. Cut the garment according to the design (Fig. 51).
3. Match all the panels for visual verification (Fig. 52).
4. Cover the raw edges of the neck using binding (Fig. 53).
5. Sew the two panels of sleeve to make one piece of sleeve and attach it into the bodice part (Fig. 54).
6. Now make strap for the frock using the wasted fabrics.
7. Hence the T-shirt is converted into kid's frock (Figs. 55 and 56).

Table 8 Size chart

Age	Height	Chest	Waist	Hip
4 years	38–42″	22″	21″	23″

Fig. 51 Cut the garment
according to design

Fig. 52 Visual verification
for matching of all panels

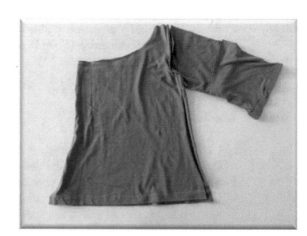

Fig. 53 Covering of raw
edges

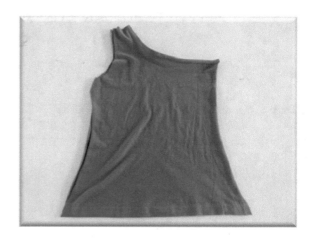

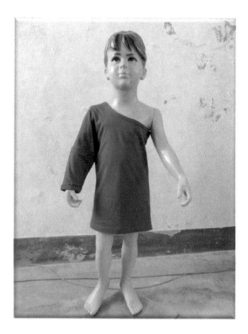

Fig. 54 Garment construction

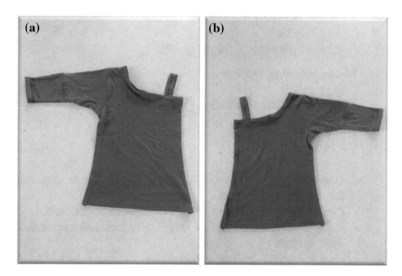

Fig. 55 Upcycled garment. **a** Front view and **b** back view

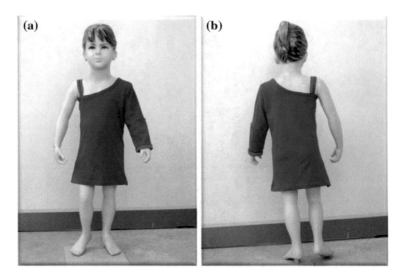

Fig. 56 Upcycled garment on dummy. **a** Front view and **b** back view

5.8 Case Study 8—Conversion of Ladies T-shirt into Kid's T-shirt

Step 1: **Selection of Garment**

- Selected garment: T-shirt (Fig. 57).

Step 2: **Identification of Defects**

- Identify the defects/damages in the garments.
- Defect: oil stain (Fig. 58).

Step 3: **Creation of Design**

- Generate ideas to upcycle the pre-consumer waste (ladies T-shirt) from various researches.
- Design those ideas into the form of illustration/sketches.
- Analyse whether the design can able to replace/alter the defects in final product.
- Sketch original product and final output (before and after).
- Design kid's T-shirt with boat neck.

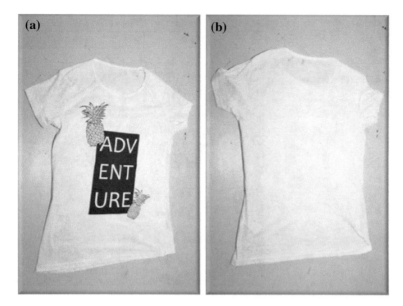

Fig. 57 Damaged/defective garments. T-shirt **a** front view and **b** back view

Fig. 58 Garment defect oil stain

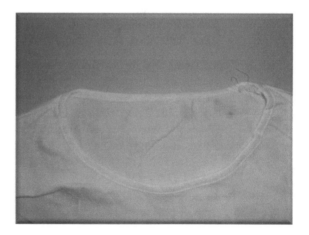

Step 4: **Evaluation of the Garment with Design**

- Select a garment from the pre-consumer waste which must be evaluated by matching the following requirements.
- The garment must able to match the sizing measurement of the final output.

Table 9 Size chart

Age	Height	Chest	Waist	Hip
4 years	38–42″	22″	21″	23″

Fig. 59 Deconstruction of garment

- Analyse whether the garment dimensions are sufficient to make alternation and construction (Keep in mind about the ease allowance, seam allowance).
- Ladies T-shirt size: medium.
- Kid's night wear size (Table 9).

Step 5:   **Construction of the Garment**

1. Remove the damaged parts in the garment (Neck area).
2. Deconstruct the garment according to the final output (Fig. 59).
3. Shape the deconstructed garment.
4. Attach the sleeve and sew the side seam.
5. Use the wasted remaining fabric from the garments to make collar for kid's T-shirt.
6. Hence T-shirt is converted into kid's T-shirt (Figs. 60 and 61).

5.9   Case Study 9—Conversion of Ladies T-shirt into Teenage T-shirt with Lace Neck

Step 1:   **Selection of Garment**

- Selected garment: T-shirt (Fig. 62).

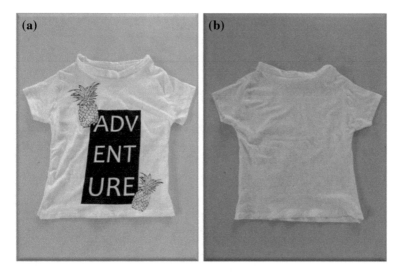

Fig. 60 Upcycled garment. **a** Front view and **b** back view

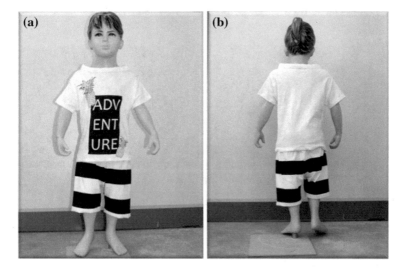

Fig. 61 Upcycled garment on the dummy. **a** Front view and **b** back view

Step 2: **Identification of Defects**

- Identify the defects/damages in the garments.
- Defects: (1) open seam and (2) misaligned (Fig. 63).

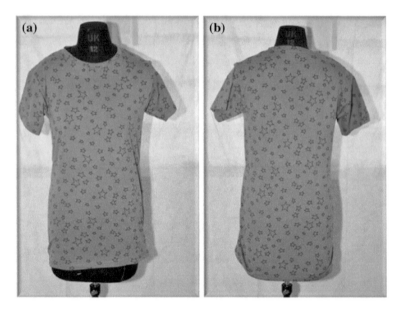

Fig. 62 Damaged/defective garment. T-shirt **a** front view and **b** back view

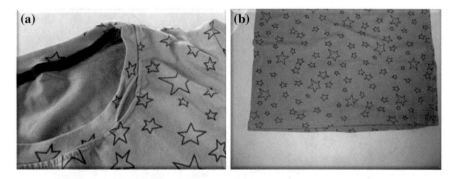

Fig. 63 Garment defects. **a** Open seam and **b** misaligned

Step 3:   **Creation of Design**

- Generate ideas to upcycle the pre-consumer waste (ladies T-shirt) from various researches.
- Design those ideas into the form of illustration/sketches.
- Analyse whether the design can able to replace/alter the defects in final product.
- Sketch original product and final output (before and after).
- Design kid's T-shirt with lace neck.

Table 10 Size chart

Age	Height	Chest	Waist	Hip
4 years	38–42″	22″	21″	23″

Step 4:   **Evaluation of the Garment with Design**

- Select a garment from the pre-consumer waste which must be evaluated by matching the following requirements.
- The garment must able to match the sizing measurement of the final output.
- Analyse whether the garment dimensions are sufficient to make alternation and construction (Keep in mind about the ease allowance, seam allowance).
- Ladies T-shirt size: medium.
- Kid's night wear size (Table 10).

Step 5:   **Construction of the Garment**

1. Deconstruct of damaged garment (Fig. 64).
2. Shape the garment according to the final output (Fig. 65).
3. Cover the raw edges using binding technique.
4. Sew the side seam.
5. Attach the laces in the neck area at both sides (Fig. 66).
6. Hence the kid's T-shirt is converted from the T-shirt (Figs. 67 and 68).

5.10   Case Study 10—Conversion of Ladies T-shirt into Kid's Night Pant

Step 1:   **Selection of Garment**

- Selected garment: T-shirt (Fig. 69).

Step 2:   **Identification of Defects**

- Identify the defects/damages in the garments.
- Defect: holes (Fig. 70).

Step 3:   **Creation of Design**

- Generate ideas to upcycle the pre-consumer waste (ladies T-shirt) from various researches.
- Design those ideas into the form of illustration/sketches.

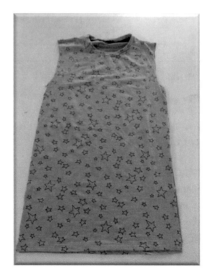

Fig. 64 Defective garment after marking as per new design

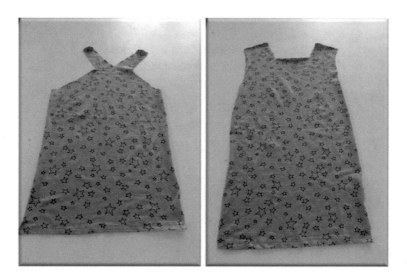

Fig. 65 Defective garment after shaping as per new design

- Analyse whether the design can able to replace/alter the defects in final product.
- Sketch original product and final output (before and after).
- Design kid's pant.

Fig. 66 Lace

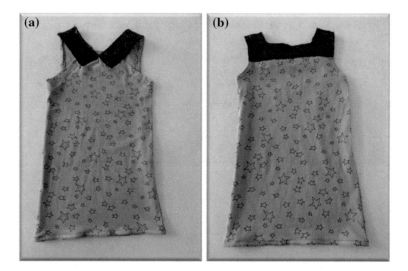

Fig. 67 Upcycled garment after attaching the lace. **a** Front view and **b** back view

Step 4: **Evaluation of the Garment with Design**

- Select a garment from the pre-consumer waste which must be evaluated by matching the following requirements.
- The garment must able to match the sizing measurement of the final output.

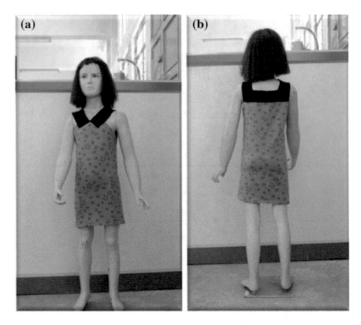

Fig. 68 Upcycled garment on the dummy. **a** Front view and **b** back view

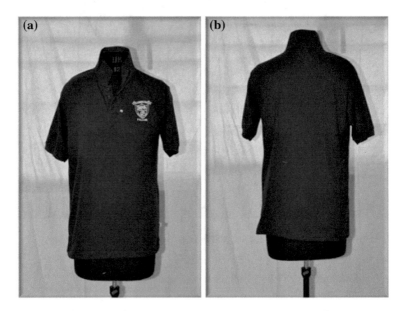

Fig. 69 Damaged/defective garment—T-shirt. **a** Front view and **b** back view

Fig. 70 Garment defect—
hole

Table 11 Size chart

Age	Height	Chest	Waist	Hip
4 years	38–42″	22″	21″	23″

- Analyse whether the garment dimensions are sufficient to make alternation and construction (Keep in mind about the ease allowance, seam allowance).
- Ladies T-shirt size: medium.
- Kid's night wear size (Table 11).

Step 5: **Construction of the Garment**

1. Deconstruct the T-shirt for upcycling (Fig. 71).
2. Use the sleeve rib for pant bottom portion (Fig. 72).
3. Joint the crotch part for both front and back panel (Fig. 73).
4. Sew the side seam (Fig. 74).
5. Sew the inseam and produce inseam.
6. Attach the pockets at the back side of the garment (Figs. 75 and 76).

Fig. 71 Defective garment
after cutting as per new design

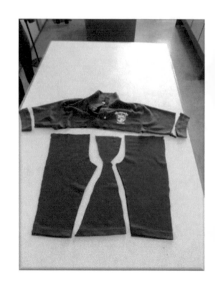

Fig. 72 Sleeve rib for
bottom portion

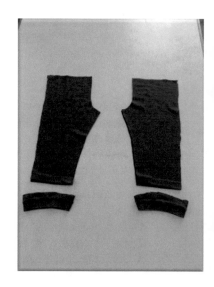

Fig. 73 Joint the crotch part
for both front and back panel

Fig. 74 Sewing side seam

Fig. 75 Attachment of pockets

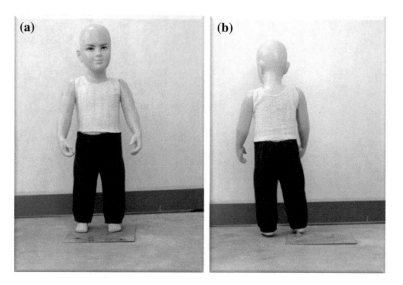

Fig. 76 Upcycled garment on the dummy. **a** Front view and **b** back view

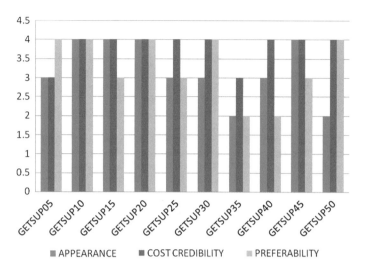

Fig. 77 Customer feedback on the upcycled garments

6 Customer Feedback

Subjective assessment has been also carried out on the upcycled garments to know the customer feedback. As the consumers are babies, parents of the babies have given their feedback on these garments. All the garments are assessed for the parameters appearance, cost credibility and preferability using 5-point Likert scale, the values being very good (5), good (4), fair (3), satisfactory (2) and poor (1). Figure 77 represents the feedback on the upcycled garments (garments are labelled as Getsup001 to Getsup010). It is evident from the graph that customers are happy with the cost credibility, appearance and preferability.

7 Conclusions

A framework for the development of upcycled garments from garment waste has been successfully presented in this chapter. The important aspect of 4R concept is "rebuy" aspect of upcycled garments. Customers have expressed positive responses on the appearance, cost credibility and preferability to the upcycled garments. In the light of above findings, the upcycled garments can be a new market that needs to be tapped at the earliest to make the world a truly sustainable. However, the challenges in the upcycling include the inherent negative association with the upcycled products, plenty of manual work and textile waste disposal systems.

References

Brooks, J. (1979). A friendly product. *New Yorker, 55*(39) 58–94. (November 12).

Fernie, J., & Azuma, N. (2004). The changing nature of Japanese fashion: Can quick response improve supply chain efficiency? *European Journal of Marketing, 38*(7), 749–769.

Hoffman, W. (2007). Logistics get trendy. *Traffic World, 271*(5), 15.

Scheirs, J. (1998). *Polymer recycling, science, technology and applications*, New York: Wiley.

The Economist. (2005). The future of fast fashion: Inditex. *The Economist, 375*(8431), 63.

UNEP. (2011). *Annual Report, 2011 through* http://www.unep.org/annualreport/2012

United Nations. (1987). *Report of the World Commission on Environment and Development Our Common Future through* http://www.un-documents.net/wced-ocf.htm

Wang, Y. (2006). *Recycling in textiles*. Cambridge, UK: Woodhead Publishing.

The Use of Recycled Fibers in Fashion and Home Products

Karen K. Leonas

Abstract As the textile, apparel, fashion, and retail industries move to become more sustainable, an area of interest is the use of recycled fiber, yarn, fabric, and product content in the development and production of new products. The decision to use recycled materials in products must occur during design and product development and continue throughout the manufacturing processes. There are several recognized stages in recycling collection, processing, and then use in a new product. Recycled materials used in textile and apparel products can be obtained throughout the textile and apparel supply chain and post-consumer collection methods. The use of recycled raw materials aligns with the larger movements of global industries toward a circular economy (vs. linear) and working to achieve a closed-loop production cycle. This chapter reviews the textile and apparel industry, factors that have influenced the generation and use of waste and recycling processes currently used today. Selected brands that have programs and products that contain recycled content are identified here.

Keywords Circular economy · Closed loop · Recycled fibers · Recycled content · Waste · Recycling · Post-consumer

1 Introduction

As the textile, apparel, fashion, and retail industries move to become more sustainable, an area of interest is the use of recycled fiber, yarn, fabric, and product content in the development and production of new products. The decision to use recycled materials in products must occur during design and product development and continue throughout manufacturing processes. There are two stages in recycling —collection and processing. Recycled materials used in textile and apparel

K.K. Leonas (✉)
Department of Textile and Apparel, Technology and Management, College of Textiles,
North Carolina State University, Raleigh, NC 27695, USA
e-mail: kleonas@ncsu.edu

© Springer Science+Business Media Singapore 2017   55
S.S. Muthu (ed.), *Textiles and Clothing Sustainability*,
Textile Science and Clothing Technology,
DOI 10.1007/978-981-10-2146-6_2

products can be obtained throughout the textile and apparel supply chain and post-consumer collection methods. The use of recycled raw materials aligns with the larger movements of global industries toward a circular economy (vs. linear) and closed-loop production. This chapter reviews the textile and apparel industry, factors that have influenced the generation and use of waste, recycling processes currently used today, and selected brands are identified in this chapter.

2 The Textile and Apparel Industry

Since the Industrial Revolution when the production of textiles became mechanized through development and commercialization of related technology including the spinning frame, power loom, and cotton gin, there had been increasing abilities to produce larger quantities of yarns and fabrics. Yarn and fabric production moved from the home and small enterprises to large, industrialized factories. In the mid- to late 1800s, the sewing process also became mechanized with the invention and patenting of the sewing machine. From 1842 to 1885, the USA issued over 7300 patents for sewing machines and accessories (Burns et al. 2011). Additional technological advancements such as motorized cutting knives and pressing equipment introduced in the late 1800s contributed to the growth of the industry and supported the development of the factory system. Although menswear was the first market segment to move to ready-to-wear apparel, other market segments including children's and women's apparels soon followed. Apparel styles for women also changed with the adoption of separates and shirtwaists, which was conducive for the ready-to-wear market.

In addition to technological developments in textile and apparel manufacturing, during the first half of the twentieth century, the development of regenerated and synthetic polymers suitable for use as textile fibers increased raw material availability. Production of manufactured fibers increased rapidly and by the 1980s the consumption of these fibers outpaced the consumption of natural fibers. Because of these many advancements, the production and availability of textiles and apparels increased dramatically. Developed countries around the world moved from prosumer to consumer societies supporting the growth of a linear economy.

Direct environmental impacts of the apparel and textile industry addressed in this chapter include the use of raw materials to produce textiles and the pollution and solid waste generated through the manufacturing process and the disposal of used textile and apparel products. In 2013, the global consumption of fibers and yarns increased to 90.1 million tons from a record of 82 million tons in 2011. To produce fibers in 2011, it required 145 million tons of coal and a couple trillion gallons of water (McGregor 2015a; Aizenshtein 2009). The US EPA estimates that textile waste occupies nearly 5 % of all landfill space and the average US citizen throws away 70 lb of clothing annually. It is also estimated that the textile recycling industry recycled approximately 3.8 billion pounds of post-consumer textile waste that accounts for approximately 15 % of the total (National Cotton Council of America 2016).

The USA is not unlike other parts of the world as these short-lived products are in either landfills or incinerators sooner than more durable ones. In 2012, in the USA alone, the incineration of synthetic fibers resulted in 1.1 MMT (million metric tons) of CO_2e (carbon dioxide equivalents) emissions, while textiles in landfills contributed a net 8.5 MMT CO_2e that year (Patagonia 2016a). Organizations throughout the world, including major apparel brands, have acknowledged the environmental impacts of the textile and apparel industries as demonstrated by the following statements:

> You cannot have infinite, unfettered growth and fast-fashion methods of consumption and production if you want to protect resources Gwen Cunningham, Circular Textile Program, (McGregor 2015a)

> The nylon and polyester polymers we use in our technical shells—which we also use in some of our other products—are neither infinite nor sustainable (Patagonia 2016a)

and

> The fashion industry is too dependent on natural resources and we must change how fashion is made. ...clothes are a necessity, However, the fashion industry requires large amounts of natural resources, lots of which can be reduced, recycled, substituted or eliminated Cecelia Brannsten, H&M (McGregor 2015a)

In January 2016, of 331 exhibitors at the Textile World, USA, there were 29 companies offering products with eco-friendly materials or using eco-friendly processes (McGregor 2016c). It is expected that increasing number of companies will begin to incorporate market sustainable products and practices at future exhibitions.

Textiles and apparel is a term that encompasses a plethora of items from the apparel worn for protection to self-expression, items in the home including linens and upholstery, geo-textiles, building materials, and automotive components to name a few. Not only does everyone use textiles in their everyday life, but in 2011 the fashion and textile manufacturing industry also employed more than a billion people globally (Hayes 2011). The movement in the textile and apparel industry from a linear to circular economy and to a closed-loop manufacturing process is seen in all product categories from activewear, basics (socks, t-shirts), fashion items to the highest performance athletic wear. Nike states '*We envision a transition from linear to circular business models and a world that demands closed-loop products - designed with better materials, made with fewer resources and assembled to allow easy reuse in new products*' (Nike 2016c). This is aligned with similar transformational changes taking place in other global industries and seems to be changing consumer's behaviors and attitudes. The goal to divert waste from the landfill is to disrupt the current production processes and waste management behaviors, both at an industry and at a personal level, that result in the reduction of waste.

3   Textile Recycling Aligned with the Circular Economy

There is movement in many industries, including the textile and apparel industry, toward the development of a circular economy and away from the traditional linear economy. Historically, the textile and apparel industry has been an excellent example of the latter. The linear economic model is represented by concepts of 'take, make, and dispose,' 'make, use, dispose,' or 'more is better.' This economy relies on large quantities of low cost, easily accessible materials. Moving to a circular economy reduces the impacts of a linear economy by creating a different system. A circular economy is known to be restorative and the concepts of 'reduce, reuse and recycle,' 'make/remake,' 'use/reuse,' and 'repurpose' represent this system. Goals of a circular economy are to use products, components, and materials to their highest value at all times. Resources are recovered and restored in the technical cycle, typically requiring human intervention. The recovered materials are then reused many times and repurposed to create a product of value (Ellen MacArthur Foundation 2012).

4   Open- and Closed-Loop Recycling

The two primary stages involved in recycling are collection and reprocessing. To create a closed-loop system, the additional stage of creating a new, recyclable product must be added. With regard to textiles and apparel industry, the collection process takes place at various points through the supply chain, and there are programs where the public can be involved in the process. Waste is also collected from sources outside the textile and apparel industry for reprocessing and use in apparel and textile end products. Reprocessing of the collected materials is critical in determining whether it will contribute to an open- or closed-loop system. In an open-loop system, the material is not recycled indefinitely and is eventually excluded from the utilization loop and diverted to the landfill. There are a number of reasons for excluding material from the loop. Two common reasons for this exclusion are: (1) degradation of the raw material that results in reduced quality and (2) incorporation of the raw material into a product that is not recyclable. In general, open-loop recycling postpones waste from being generated but does not ultimately keep product from the waste stream. An alternate and more sustainable strategy is the closed-loop recycling. The recycling of the material is indefinite and without degradation. This conversion of the used product back to raw materials allows for making the same product over and over again. Biodegradable products are also a part of the closed-loop recycling system. This is also known as cradle-to-cradle (Payne 2015). Payne (2015) provides a thorough overview of open- and closed-loop recycling for textile and apparel products including schematics of open- and closed-loop recycling systems (Figs. 1 and 2).

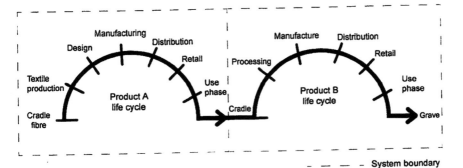

Fig. 1 Open loop recycling from Payne (2015, p. 107)

Fig. 2 Cradle-to-cradle closed-loop recycling from Payne (2015, p. 113)

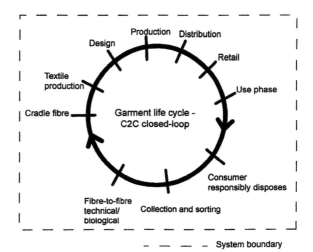

A number of textile and apparel companies are working to achieve closed-loop recycling including Patagonia as seen on the company's Web site *'best option at this point is to buy recycled-content fabrics and to continually find new ways to recycle our products at the end of their life and keep them out of traditional disposal methods'* (Patagonia 2016a). As a brand, Patagonia is aware of challenges with the use of recycled fibers. As many of their products fall into the high-performance category, they are concerned with the loss of durability in the resulting fabric. High-quality fabrics that exceed the expected performance have been developed from the recycled polyester. However, they have found that recycled nylon is less durable, more difficult to obtain the necessary quantity, and has increased weight when compared to virgin nylon (Patagonia 2016a).

5 Recycling and Reuse

Recycling and reuse of materials is not new to the textile and apparel industry. Payne distinguishes between recycling and reuse as follows: Recycling *'refers to the breakdown of product into its raw materials in order for the raw material to be reclaimed and used in new products. In contrast, reuse refers to an existing product being used again within the same production chain. Textile recycling may involve reclaiming pre- consumer waste or post-consumer waste'* (Payne 2015, p. 105). For centuries, end products were repurposed after they have reached the end of the their use in one product.

There are a variety of methods in which reuse occurs. The used product can be disassembled and then reassembled into a new, and possibly different, product. Examples of repurposing are frequently seen in pop culture including a scene from *Gone with the Wind*, where Mamie removed the green velvet drapes and repurposes them into a gown for Scarlett O'Hara (Mitchell 1936). Historically, it was common that when apparel was no longer useful, once outgrown or no longer in style, it was remade to fit someone else or redesigned to create a more stylish garment. The recycling of wool is hundreds of years old. After apparel (i.e., wool sweaters) had been worn threadbare, it was collected and shredded into individual fibers and then converted into blankets. Today it is quite common for apparel items to be donated to charities for resale or discarded in the trash bin after they have fulfilled their initial use. In some cases, items that are no longer useful as their original product are used for other purposes such as rags or stuffing. Hawley provides an extensive review of the many ways in which discarded apparel is reused. There is a detailed schematic (Fig. 3) of the multiple options for post-consumer textile products included in her work (Hawley 2015).

Recycling is the breakdown of a product into its raw materials. For centuries, textile products (apparel and fabrics) were broken down to the yarn stage and the yarn was used to produce different knitted or woven fabrics. In some cases, the yarns are further broken down to the fiber stage and then the fibers were respun into yarns to be used in new textile products. This was quite common prior to the mid-twentieth century. In 1939, the US Federal Trade Commission introduced the Wool Products Labeling Act which requires accurate labeling of wool products that distinguishes between fibers that have never been reclaimed from woven or felted wool products, identified as 'wool' and 'recycled wool' in which the fiber used has been previously been spun, woven, knitted, or felted into a wool product (Federal Trade Commission 2016).

With the introduction of manufactured fibers in the late nineteenth century and synthetic fibers in the twentieth century, the practice of breaking down post-consumer products and reusing the fabric, yarns, and/or fibers was reduced for several reasons. With increased raw materials available, products were discarded rather than reused or recycled. Next, as natural and man-made fibers were blended and products with mixed fibers became more popular, it was difficult to separate the fibers by generic class. This separation is critical in the recycling process due to the

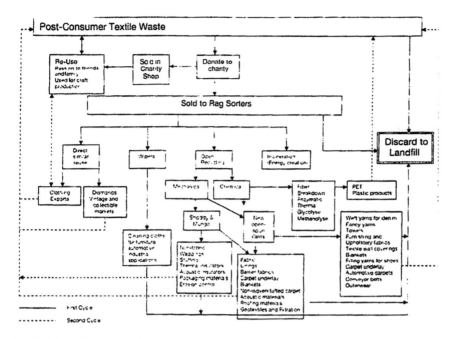

Fig. 3 Schematic showing the multiple options for post-consumer textile products from Hawley (2015, p. 212)

varied processing and performance parameters required by each individual fiber type. A challenge facing the industry related to the recycling process is the separation of the fibers by chemical class. There are a number of processes where non-contaminated waste (pure generic fiber content) can be collected throughout the supply chain. In addition, technology is being developed to achieve the separation of fibers. Today, mixed fiber and/or contaminated waste are commonly used as a fuel through incineration. However, technological development shows promise for increased success in collection, separation, and use of recycled mixed fiber materials.

Textile wastes can be classified into several ways. Commonly they are classified as to the point in which the supply chain they are collected. These include three categories: (1) pre-consumer waste, (2) post-industrial textile waste, and (3) post-consumer textile waste. There are also four recycling approaches identified in the literature. The first refers to the collection process; the second, third, and fourth focus on the processing of the waste. These approaches are: (1) primary—recycling industrial scrap; (2) secondary—processing a post-consumer product into raw materials; (3) tertiary—converting plastic wastes into basic chemical monomers called fuels; and (4) quaternary—incinerating waste as a way of reclaiming the embedded energy (Vadicherla and Saravanan 2014).

Wastes generated by the original manufacturer that never reaches the consumers are best classified as pre-consumer waste. Post-industrial textile wastes are generated during the manufacturing process. On average, about 15 % of fabric used in garment production is cut, discarded, and wasted in the process, which contributes to post-industrial waste (Beitch 2015).

Post-consumer textile wastes are the wastes that come from the consumer. Collection at this point in the supply chain requires buy-in of the public and is recovered from the consumer supply chain (Vadicherla and Saravanan 2014).

Although over the past quarter century, recycling has increasingly become an intrinsic part of our everyday language, the textiles and apparel industry has been slow to adopt this practice. In the USA alone, 14.3 million tons of textile waste were created last year. While 2.3 million tons of it was recycled, the goal is to increase these numbers. The textile and apparel industry is global and uses a resource-intensive supply chain that causes massive waste and creates environmental harm. The industry is one of the world's largest producers of toxic environmental waste, affecting air, water, and soil resources. This poses major challenges for brands and their efforts to produce eco-friendly products (Evrnu 2015a, b).

Preferences for recycling of textile wastes in the industry appear to the predominantly thermoplastic polymer-based fibers due to the ease and feasibility of reprocessing them. In addition, these materials have the ability to take on different forms and shapes after recycling. Natural fibers such as cotton, wool, and silk are also finding their ways into recycling streams. The majority of textile wastes reported in the literature included polyester, polyethylene, nylon, p-aramid, carbon, silk, polybutylene terephthalate, bamboo, cotton, and kenaf (Vadicherla and Saravanan 2014).

Using recycled thermoplastic fibers reduces the dependence on petroleum as a source of raw materials. It curbs discards, thereby prolonging landfill life and reducing toxic emissions from incinerators. It helps to promote new recycling streams for polyester clothing that is no longer wearable.

6 The Textile and Apparel Supply Chain

It is important to review the textile and apparel supply chain to better understand the recycling approaches and categories described in the previous section. Recycling involves two primary steps, collection and processing. To achieve a closed-loop system, the processed waste must then be used in new recyclable products. The collection of waste can occur at various points throughout the textile and apparel supply chain. The textile, apparel, and fashion industry is complex and includes a variety of product categories covering diverse market sectors. Generally, it is accepted that production of fashion and textiles utilizes one of the longest and most complex industrial chains in the manufacturing industry (Hayes 2011).

Due to the diversity of raw materials and end products, there are numerous and varying supply chains used. A simplified supply chain for fashion, textile, apparel, and retail items will include raw materials (fibers and yarns), fabric production, end product manufacturing, the retailer, and the consumer (Kincade and Gibson 2010).

Fig. 4 Textile product supply chain

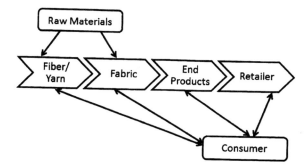

See Fig. 4. Supply chains are based on the specific product and expand to include all findings, finishes, etc., used in the final product.

Pre-consumer wastes are generated throughout the first stages of the supply chain. In the raw materials sector (fiber and yarn production), ginning wastes, opening wastes, carding wastes, comber noils, combed waste yarns, roving wastes, ring-spinning waste fibers, ring-spun waste yarns, open-end spinning waste fibers, and open-end spinning yarn wastes are commonly collected for recycling. From sector two (fabric production), knitting waste yarns, woven fabric scraps, trimmed selvedges, poor quality fabrics, and warper tie in scrap yarns are examples of wastes collected and suitable for recycling. Examples of post-industrial waste can be found at several points within the supply chain. When producing thread, it is common to collect waste thread throughtout the production process. American and Efird thread generates approximately 20,000 lb of waste thread each month that is diverted from the landfill through their recycling programs (Summers 2016). In sector three, end product manufacturing, which includes cut-and-sew operations, there is significant fabric waste resulting from leftover fabric between individual pattern pieces and areas along the selvage (Payne 2015). Scrap fabrics from the wet processing sector of the industry also yields suitable waste for recycling (Vadicherla and Saravanan 2014).

Post-consumer waste is received from the public, which includes items that have no more use for the owner. This commonly includes donated and discarded apparel and some plastic items such as plastic bottles made from polyethylene terephthalate. Nylon can also be recycled and large source of post-consumer nylon waste is fishing nets left in the ocean.

As noted previously, the second major step in recycling is the processing. This is dependent on the chemical and physical characteristics of the collected items and the demand for the recycled product. The chemical properties are determined by the fiber.

There are numerous of fibers used in apparel, and home textiles are selected based on a variety of considerations including comfort, flexibility, appearance, cost, and hand with attention to other properties as determined by the function necessary in the product. The selection of the appropriate fiber begins during product development and throughout the design process. Fibers are the building block of the textiles with their performance properties determined by their chemical and physical properties. Fibers are classified as natural or manufactured depending on how they are produced. Natural fibers commonly used in apparel and home textile products include cotton, flax, hemp, wool, and silk. Manufactured fibers are further categorized as those regenerated from natural materials (regenerated) such as cellulosic and protein, or from a petrochemical base (synthetic). Regenerated fibers used in textiles include rayon, bamboo, and Lyocell, and general use of synthetic fibers include polyester, nylon, spandex, and acrylic.

Polyester is now the single largest fiber group in the world. In 2004, world demand for polyester was 24.7 million tons, slightly ahead of the demand for raw cotton (Hayes 2011). Cotton makes up 35.3 % of fiber used in apparel products (National Cotton Council of America 2016). Large percentages of cotton and polyester used in textiles and apparel products contribute to the interest in the development of successful recycling programs. Benefits of polyester recycling methods include reducing the high ecological and social cost of oil reducing petrochemical pollution and reduction production emissions, including environmentally damaging chemical such as cobalt, manganese salts, sodium bromide, antimony oxide, and titanium dioxide (Hayes 2011).

Studies have shown that recycled polyester fibers, yarns, and fabrics made of different textiles and polymeric bottle wastes possess physical properties similar to that of virgin polyester.

7   Recycling in the Textile and Apparel Industry

In the textile and apparel industry, there are numerous examples of recycling collection and processing programs. There has been great effort and interest in recycling of polyester and cotton due to their wide use. The most successful programs focus on polyester and cotton. However, other fibers can be recycled, and nylon and wool have successful recycling programs. In addition, some firms are beginning to recycle aramids such as Kevlar® (National Spinning 2016). In the processing, there are two categories commonly used in textiles, mechanical and chemical.

8   Mechanical Recycling

Mechanical recycling processes can result in the production of fabric, yarns, or fibers to be used in new products. The discarded textile is opened up, apparel is disassembled, and fabrics are cut into smaller pieces. It is then passed through a rotating drum to continue the breakdown and fibers are obtained. This process is known as garneting. The resulting fiber characteristics of length, fineness, strength, polymer, and color determine the quality and what the most appropriate new end product would be. Typically, waste collected from the manufacturing supply chain will produce higher-quality recycled fibers that those collected from post-consumer waste. The pre-consumer and post-industrial processed waste can be respun into yarns which woven or knitted into fabrics, and then used in apparel, sheeting, and upholstery. Medium-grade fibers can be used to make fabrics but are used in end products such as wipes and fillings. Lower-quality fibers will be used as reinforcement in other structures (i.e., concrete), nonwoven fabrics, carpet underlays, shoe inlays, automotive sound and thermal insulation, home insulation, stuffing for toys, and other end products.

Plastics, including plastic bottles and thermoplastic fibers, are commonly recycled using mechanical methods. In these cases, the plastic waste is chopped into small flakes that are melted and then extruded into a form to be used in a new product. This melt can be extruded into filaments, yarns, or other formed products. There is a little noticeable difference between virgin polyester and recycled polyester fibers. This is a common method for reprocessing the plastic water bottles and fishing nets. However, not all recycled thermoplastic fibers have properties similar to virgin fibers.

9 Chemical Recycling

Chemical recycling is the other method commonly used to process the collected waste in the textile industry. Synthetic fibers including polyesters, polyamides, and polyolefins can be chemically recycled. This falls under the tertiary class of recycling which requires the breaking down of the synthetic fibers for repolymerization. This process can be used when PET plastic water bottles are recycled. Whether it is the collection of used polyester apparel, fabric scraps, yarns waste, or other plastics, the recycled items are broken into small pieces from which chips are produced. The chips are decomposed to form dimethyl terephthalate, which is then repolymerized and spun into new polyester fibers, filaments, and yarns.

Blends are in particular challenging to recycle due to the disparate physical and chemical properties of the fibers in the waste. Cotton and polyester blends are one of the most commonly used apparel and home textile items. Chemical recycling has proven successful when used with blended materials as it uses a selective degradation method. In products of cotton and polyester, the fibers can be chemically separated and then reformed into new fibers. Currently, there is a process being developed using n-methylmorpholine-N-Oxide, which dissolves cellulose. The dissolved cellulose and polyester are separated by filtration and the captured polyester is respun into a fiber, filament, or yarn. The dissolved cellulose can be used in the production of regenerated cellulosic fibers including Lyocell (McGregor 2015b; Zamani 2011).

Nylon and spandex is a blend commonly found in high-performance sportswear and activewear. Generally, the percentage of nylon is much greater than that of spandex and nylon can be recycled and reused. It is known that spandex can be removed from blended fabrics by dissolving it in solvents such as N, N-dimethylformamide. However, this solvent is expensive and there are environmental concerns with its use. There has been success by first treating the blended fabric with heat to degrade the spandex, and then exposing the fabric to a washing process using ethanol, which effectively removes the spandex residue leaving only the nylon (Yin et al. 2013).

Today, for products of single fiber content fabrics, mechanical recycling is more prominent. The chemical recycling procedures require more energy consumption and there is high capital investment so this option is only practical for large-scale manufacturers. As the technology improves, the demand for recycled content increases, and as the cost of virgin raw materials increases, there is likely to be a shift from mechanical to chemical recycling of these materials (Agrawal et al. 2015).

10 Textile and Apparel Recycling Programs in the Industry

Throughout the textile and apparel industry, there are numerous programs that focus on collecting and processing of textile waste. Several companies have established used apparel and or footwear drop boxes where the product goes through a recycling process and the materials can be reused. There are few closed-loop programs but an increasing number of companies are have set closed-loop production as a goal. In addition, there are programs related to brands promoting the use of recycled content. As companies begin to consider their role with regard to environmental impact of their products, there are many considerations. The information here focuses primarily on collection, processing, and reuse of recycled content in products throughout the supply chain including products sold directly to the consumer. Although not all programs are included, a variety of supply chain sectors, product categories, and retailers are highlighted. The decision to use recycled content must begin with the product development and design team and then be implemented through the sourcing of certified recycled fibers, yarns, and/or fabrics. Challenges when using recycled content include an increased cost due to the additional procession costs, limited color selection, consumer acceptance, less uniform fibers associated with mechanical recycled natural fibers, which lead to production difficulties, and an uncertain supply chain.

As the industry of the industry to a more closed-loop manufacturing model efforts to collect and process waste to be used in new products is increasing. In this chapter, recycling in the textile, apparel, and fashion industry will be explored focusing on examples of brands that are utilizing recycled content in their products.

11 Raw Materials—Fibers/Yarns

The use of recycled fibers to produce yarns that are then used in final products can be found from home textiles, to fashion items, to sportswear, and many other product categories. Cotton and polyester are probably the most common fibers recycled, but other fibers including wool, nylon, and even aramids are being recycled in yarn production. There are many yarn manufacturers incorporating recycled content into their products. Much of the research and development revolving around this takes place in this sector (raw material) of the industry supply chain. This section will highlight several yarn manufactures and their recycling programs.

12 Unifi[1]

Repreve® is a brand of recycled fiber that made from recycled polyester including post-consumer plastic bottles and post-industrial waste from manufacturing wastes. Using post-consumer waste offsets the need to use new resources (i.e., petroleum) and therefore, there is a reduction in the production of greenhouse gasses. Produced by Unifi, Repreve® is used in numerous brands including Quicksilver, Haggar Clothing, Patagonia, Roxy, Katmandu, Russell, Starter, Adidas, to name a few (Repreve 2016). Through this process, over 630 million plastic bottles have diverted from the landfill and used in fibers. There are several programs within the Repreve® brand including the Repreve® Textile Takeback Program and Repreve 100. In 2015, The Repreve Textile Takeback program, with the help of companies they partner with like The North Face, surpassed three million pounds of takeback fabric and have expanded the program into other categories including apparel, automotive, hospitality, healthcare, and contract furnishings (Beitch 2015). Jay Hertwig, Vice President of global brand sales and marketing for Unifi made the following statement: '*At Unifi, we continue to expand the process for making Repreve, engineering new ways to recycle materials throughout the supply chain. We are proud to provide our customers with sustainable solutions for recycling their own waste into new products, whether it's bottles or fabric scraps.*'

13 Tenjin

Ecocircle® is a fiber produced from recycled polyester. The process is a fiber-to-fiber polyester recycling system that was developed by Teijin Fibers. It is a closed-loop recycling system for polyester products and a chemical recycling processing is used. The fabrics produced from Ecocircle® are innovative and developed for use in the men's and women's active wear markets. In 2002 when the closed-loop recycling system began, only three companies were involved. By 2016, 150 companies are participating (Tenjin 2016).

---

[1]"Unifi extends recycled fibre range, moves production to China and introduces verification programme", 2009, *Advances in Textile Technology.*

14 Aquafil

Nylon 6 can also be recycled and the Econyl® Regeneration System was introduced in 2011 and provides opportunities for a new supply chain, which is infinite, innovative, and sustainable. In this system, the Nylon 6 polymers are produced using both post-consumer waste and preconsumer waste. Aquafil began working on the process in 1998 and they continue to expand the process by increasing the percentage of post-consumer waste collection sites for the program. Much of the post-consumer waste comes from fishing nets discarded in the ocean and the pile of used carpet (carpet fluff). They continue to increase the waste collection network and collect materials for recycling throughout the world. Currently, they have collection sites in the USA, Egypt, Pakistan, Thailand, Norway, and Turkey. Their collection strategy includes having partnerships with institutions, customers, and various consortia including Carpet America Recovery Effort. They have partnerships with brands including Levi's, Milliken, and Speedo who use the yarn in their products (Aquafil Global 2016; Econyl® 2016; McGregor 2015c).

15 Martex Fiber

Martex Fiber collects textile waste and supplies the textile and related industry with recycled cotton textiles. They use a 360° recycling process and provide goods for use in a variety of industries including automotive, bedding, home furnishings, construction, nonwovens, and geo-textiles (Martex Reclaimed Fiber 2016)

Jimtex yarns, a division of Martex Fiber, specializes in the production of spun yarns using reclaimed cotton fibers that can be found in home textiles, apparel, or hosiery. Their ECO2Cotton® is ecological and economical and produced from post-industrial waste obtained from cut-and-sew operations in apparel manufacturing. The process first requires that the post-industrial waste is sorted by color and then is defiberized. The reclaimed cotton fibers are blended with acrylic or polyester for strength. Jimtex provides yarns with 70–75 % cotton content, in a variety of colors, yarn sizes, and plys (JimTex Yarns 2014).

16 Evrnu

Evrnu uses post-consumer cotton garment waste to create a high-quality, bio-based fiber (Enrvu 2015a). After collection of the cotton garment waste, the dyes and other contaminates are removed. The cotton is then pulped and broken down into the fiber molecules. The molecules are then recombined and extruded as a new fiber. It is possible to engineer certain characteristics of the new fiber including the diameter and cross-sectional shape. The properties include a filament that is finer than silk and stronger than cotton (Enrvu 2015b).

17 EcoAlf

EcoAlf is an apparel brand that was founded in 2009 with the goal of creating fashion that is eco-friendly and sustainable. In the fall of 2015, they launched a project to use plastic waste from the ocean by collecting trash from the seabed and then reprocessing it into yarns used in fabrics. They have recycled fishing nets, plastic bottles, tires, and other wastes into jackets, shoes, and bags. To collect the waste, they have collaborated with fishing vessels in the Mediterranean Sea to 'catch' the plastic waste (Advanced Textiles Source 2013).

In addition to waste found in the oceans, they collect other post-consumer waste, such as coffee and cotton, and post-industrial wool. EcoAlf competes in the mid- to high-end fashion market with three lines of clothing and accessories. In addition to their own retail stores, their products are sold over 300 other sites including Nordstrom, Barney's, Urban Outfitters, Saks Fifth Avenue, Harrods, Goop, and Bloomingdales.

18 Timberland

In 2012, it was reported Camtex Fabrics, the maker of Cambrelle® shoe linings introduced fabrics with recycled content. This was at the request of Timberland for a product in their Earthkeeper shoes, boots, and clothing. The goal at that time was to develop a material that had at least 50 % recycled content. The product was polyester and the recycled content was from recycled bottles (Camtex Fabrics Ltd. 2015). It was reported in April 2015 that Timberland increased the use of renewable, organic, and recycled (ROR) materials in its footwear incorporating ROR in 79 % of its offering in 2014, a 9 % increase from 2013. More than 1.25 million pounds of recycled PET was used in the branded shoes and 6.9 million featured outsoles containing up to 42 % recycled rubber. The timberland apparel lines contained 36.7 % ROR in 2013 but that dropped to 18.8 % of all materials in 2014. The primary challenge was due to cost (McGregor 2015d).

19 Nike

Nike, like The North Face, views sustainability hand in hand with innovation. For a number of years, Nike has been looking for ways to reduce their environmental impact. Beginning in the early 1990s, the Reuse-A-Shoe program was introduced.

In this program, post-consumer wastes (in the form of old athletic shoes) are processed into Nike Grind and then reused in new products (Nike 2016b). In February 2010, Nike announced its fabric suppliers would source discarded plastic (PET) bottles from Japanese and Taiwanese landfill sites to produce new yarn for use in national soccer team jerseys. The jerseys for Brazil, the Netherlands, Portugal, Serbia, and Slovenia were made completely from recycled polyester. It was estimated that this project alone diverted 13 million plastic bottles from the landfill (Hayes 2011).

In most recent sustainability report that Nike released, it was reported that they continue to reduce their environmental footprint throughout the product lifecycle; one way of doing this is to close the production loop. About 60 % of the environmental impact in a pair of its shoes is embedded in the materials used. Approximately and 71 % of Nike's footwear and apparel products contain recycled content and can be found in everything from trims to the flyknit shoes (McGregor 2016b). In FY15, 54 million pounds of post-industrial waste, factory scrap, were reprocessed and then used in new Nike footwear and apparel products. Nike identifies their Flyknit shoes and Nike Grind programs as their two most sustainable innovations (Nike 2016a).

20 Speedo

Speedo, a PVH Corp licensee, launched Powerflex Eco® in August 2015. This was a partnership with Aquafil, and Italian yarn maker. The partnership involves the collection of fabric scraps from cut-and-sew manufacturers which was then processed into the fiber Econyl®. Econyl® is a synthetic textile made using a variety of wastes including the post-industrial scraps and post-consumer waste such waste abandoned fishing nets and old carpets. Powerflex Eco® is a combination of Econyl® (78 %) and extra-life nylon fabric (22 %) with chlorine-resistant pieces that retain their shape up to 10 times longer than traditional swimwear. The collection has a price range of 40–79 UDS which is like that of other similar speedo products (McGregor 2015c).

21 Adidas

In 2014, the German sportswear manufacturer announced a partnership with Parley for the Oceans, an organization that is working to eliminate plastic waste in the ocean. Initial products that Adidas created from this post-consumer waste included 3D-printed running sneakers and apparel using recycled content. Adidas developed a shoe with the upper made from yarns and filaments recycled from ocean waste (Shepherd 2016; Velasquez 2014a).

In 2015, Adidas announced a program known as Sports Infinity where the material is a 3D material that can be recycled over and over. Sports Infinity is a partnership that fosters working relationships among industry and academic experts. It is led by Adidas, funded by the European Commission, and partners include BASF, KISKA, the Center of Technical Textiles at the University of Leeds, Fredrich-Alexander Universitat Erlangen-Numberg in Germany and others. The Sports Infinity program will promote development of innovative processes and products that will move the industry closer to closed-loop recycling programs (Lamicella 2015).

22 Hanes

In 2010, Hanes introduced a new line, EcoSmart®. This line includes apparel items with recycled cotton and/or polyester fiber content. Products include fleece apparel, socks, polo's and t-shirts and items are available for men, women, and girls (Sustainable Brands 2010; Hanes for Good 2016). The Hanes EcoSmart® and Champion Future Friendly® apparel reuses polyester from recycled plastic bottles. Hanes owns approximately 80 % of their own production plants and from these they collect post-industrial wastes from the cut-and-sew operations. In a partnership with Martex Fiber, Jimtex Yarns division, they send the collected post-industrial waste to the yarn production plant in Lincolnton Georgia and then recused in the EcoSmart® products. The Black EcoSmart socks are made with 55 % recycled cotton fiber, which is all of the cotton used in these socks (Hanes for Good 2016).

23 H&M

H&M, the Sweedish-based fast-fashion apparel retailer, introduced two eco-collections in 2014, Conscious Collection and Conscious Exclusive. The Conscious lines are designed with more sustainable materials including organic cotton, Tencel®, hemp, and recycled components. One purpose of these lines was to show how eco-friendly garments can also be stylish and fashionable. In 2016, 2 years after the launch of the Conscious Lines, organic, recycled and Better Cotton represented 31 % of total cotton use. H&M claims to be one the biggest users of recycled polyester in the world (McGregor 2016a).

In 2013, H&M launched its in-store Garment Collecting Initiative and by 2015 they had collected 19,000 tons of discarded clothing. In 2014, they released the first products from a closed-loop system. These items were made with 20 % of recycled materials. Also in 2014, 10 new denim styles were released that contained recycled cotton fiber from the clothing collected. This supported their goal to move from linear production model to a circular one by closing the loop for textiles (McGregor 2016a).

H&M was selected to design and provide a wide range of uniforms for their home country's athletes for the 2016 Summer Olympics and Para Olympics. Items included outfits for the opening ceremonies, closing ceremonies, and athletic competitions in between and included recycle content. Pernilla Wohlfahrt, the design and creative director at H&M stated: '*We are truly honored to also do the competition outfits for some selected sports. The result is a technical, high-fashion Olympic collection with a lot of the garments made in sustainable materials such as recycled polyester*' (Porter 2016).

24 The North Face

The North Face, a brand that is a part of VF Corporation, is a leading supplier of technically innovative outdoor gear. The North Face development team knows there is a deep connection between sustainability and innovation. A goal in improving the sustainability of their products is to use recycled materials whenever possible. In June 2011, about 15 % of total material volume was from recycled content, accounting for about $150 million of product sold (Moore 2011).

In 2015, The North Face announced it would incorporate Unifi, Inc.'s Repreve® into its Denali line of fleece jackets. Three eco-friendly materials were integrated into the Denali jackets: (1) Repreve® recycled yarn, Repreve® WaterWise yarn with color technology, and Repreve® Textile Takeback yarn reprocessed from leftover fabric and recycled plastic bottles.

By using Repreve® fleece products, over 30 million plastic bottles are diverted from landfills and used to create Denali jackets each year. Repreve's WaterWise yarn with color technology also reduces the amount of water and chemicals needed to dye the fabric. The Denali jackets are available in black and heather gray reducing the amount of water, chemicals, and energy required in the fabric dyeing and finishing process.

Unifi and The North Face have developed another level of collaboration by collection of the post-industrial waste created during the production of the Denali Jackets. This waste is sent to Unifi's Repreve® recycling center to be processed into the Repreve® Takeback yarn. The yarn is then knitted into new Denali Jackets and so they have achieved a closed-loop system.

For every 10 Denali jackets produced, a sufficient amount of fabric scrap is collected to produce an additional four jackets.

25 Patagonia

Patagonia has several established recycling programs and actively uses recycled fiber in their products. The recycling programs are in place used for polyester, wool, and cotton. Patagonia began making recycled polyester from plastic soda bottles in 1993 and was the first outdoor clothing manufacturer to transform trash into fleece. They started using fiber-to-fiber recycling system to keep used clothing products out of the waste stream and trash incinerators. Today, they not only used soda and water bottles, but they also collect post-industrial manufacturing waste and post-consumer worn-out garments for reprocessing and use in new apparel. They also have increased the number of products that have recycled polyester content including Capilene® baselayers, shell jackets, board shorts, and fleece. Using recycled polyester lessens the dependence on petroleum as a source of raw materials (Patagonia 2016c).

Patagonia also uses recycled wool in its wool products. To control quality, a meticulous sorting process is necessary. The materials are sorted by color prior to shredding. By selecting and blending colors of dyed wool fabrics and garments, the dyeing process would be eliminated further reducing energy, water, and chemicals (Patagonia 2016d).

Patagonia has a partnership with the TAL Group, a large garment manufacturer, where the post-industrial cotton scraps from their China and Malaysia factories are collected and then reprocessed. This cutting-room scrap garneted to fiberize it, spun into yarns that are woven or knitted into fabric, which is then used in the production of new products. The scraps from 16 virgin cotton shirts can be turned into one cotton shirt produced from reclaimed fibers.

In reality, the reclaimed fiber is combined with virgin organic cotton and used in the Men's Reclaimed Cotton Hoody and Women's Reclaimed Cotton Crew (Patagonia 2016b).

Patagonia is moving toward more and more product with recycled content. On their Web site, under the heading 'What do we Think,' it states that the *'best option at this point is to buy recycled-content fabrics and to continually find new ways to recycle our products at the end of their life and keep them out of traditional disposal methods'* (Patagonia 2016a).

26 Cone Denim

A partnership between Cone Denim and Unifi in 2014 resulted in the development and launch of a soft stretch denim, Cone Touch™. The denim is designed to provide better comfort and stretch for jeans. ConeTouch™ incorporates Unifi's Repreve® post-consumer recycled polyester content fibers and yarns. Each pair of jeans made with ConeTouch™ contains an average of eight recycled bottles (Velasquez 2014b). The Cone Touch™ products add to Cone's Sustainable™ line of eco-friendly products.

27 Levi Straus & Co

Levi Strauss & Company has several programs that incorporate recycled fibers into their products. In Levi's (2012) introduced a new collection of denim identified as Waste Less™. Each product contains a minimum of 20 % post-consumer waste.

The waste included PET plastic bottles and black food trays. They collaborated with municipal recycling programs throughout the USA to collect the waste. The Waste Less jeans are available for both men and women and each pair will have, on average, eight to 12 plastic bottles. The post-consumer waste is first sorted by color, crushed into flakes, melted, and extruded as a fiber (Green Retail Decisions 2012; Levi's 2012).

In Spring 2015, Levi Strauss & Co. and Evrnu, SPC announced that they have created a jean made from regenerated post-consumer cotton waste. The process uses a new, patent-pending recycling technology to turn discarded consumer waste into a renewable fiber. The jean is made from approximately fiber-discarded t-shirts and virgin cotton. The warp yarns are virgin cotton and the new fiber produced by Evrnu is in the filling direction. Currently, there are several prototypes of the garment and it is hoped that the new technology can be put in place on a large scale to meet the industry demand (Peters 2016).

28 Conclusion

The textile and apparel industry is moving toward a circular economy and closed-loop manufacturing. To achieve this, one feasible method is the use of recycled materials in textile and apparel products. The future of recycling relies heavily on the development of new advanced technologies and approaches for material processing (without quality loss), collection, sorting, processing, and utilization in a new product that is also recyclable. Creating a demand for new products with recycled content is critical. It is important to include the recycled content in the design and product development stages of fashion and home products but there is also a need to encourage flows that promote recycling and reuse.

Sarah Ditty, Deputy Editor of Source Intelligence, believes there are four reasons that companies are using recycled materials: (1) driven by media scrutiny; (2) consumer demands; (3) cost; and (4) resource scarcity. '*Big companies know they have to be inventive and innovative to survive, but their supply chains are so complex it takes a lot of time and money to implement new systems,*' she says. '*It's slow moving and a long journey, but we're on the right tracks*' (Rivera 2013).

There are many challenges and opportunities facing the industry with regard to environmental sustainability. Partnerships and collaboration will be critical to successfully addressing these developing efficient, cost–effective, and closed-loop recycling programs. Levi Strauss believes '*competition and collaboration must go hand in hand to push progress in the industry*' (Sustainable Apparel Collation 2016).

References

Advanced Textiles Source. (2013). *Ecoalf upcycles ocean trash into yarns for fabric*. Available at http://advancedtextilessource.com/2016/02/ecoalf-upcycles-ocean-trash-into-yarns-for-fabric/. (18 May 2016).

Agrawal, Y., Kapoor, R., Malik, T. & Raghuwanshi, V. (2015). *Recycling of plastic bottles into yarn & fabric*. Available at http://www.textilevaluechain.com/index.php/article/technical/item/247-recycling-of-plastic-bottles-into-yarn-fabric. (12 May 2016).

Aizenshtein, E. (2009). Polyester fibres continue to dominate on the world textile raw materials balance sheet. *Fibre Chemistry, 41*(1), Springer Science+Business Media, Inc.

Aquafil Global. (2016). *The Econyl® project*. Available at http://www.aquafil.com/sustainability/the-econyl-project/. (21 May 2016).

Beitch, S. (2015). *The North face incorporates REPREVE Technology into Denali Jacket* [Online]. Available at https://sourcingjournalonline-com.prox.lib.ncsu.edu/north-face-incorporates-repreve-technology-denali-jacket-sb/. Accessed April 26, 2016.

Burns, L. D., Mullet, K. K., & Bryant, N. O. (2011). *The business of fashion: Designing, manufacturing, and marketing* (4th ed.). New York: Fairchild Books.

Camtex Fabrics Ltd. (2015). *Cambrelle® boosts Timberland's green credentials*. Available at http://www.cambrelle.com/NewsItem/?id=126420. (13 May 2016).

Econyl®. (2016). *Aquafil partners with Levi Strauss & Co. to produce sustainable jeans* [Online]. Available at http://www.econyl.com/press/aquafil-partners-with-levi-strauss-co-to-produce-sustainable-jeans/. (15 May 2016).

Ellen MacArthur Foundation. (2012). Towards the circular economy: economic and business rationale for an accelerated transition. Available from https://www.ellenmacarthurfoundation.org/circular-economy/overview/concept. (20 April 2016).

Evrnu. (2015a). *The Evrnu technology* [Online]. Available at http://www.evrnu.com/technology/. (06 May 2016).

Evrnu. (2015b). *The future of apparel* [Online]. Evrnu home page, Available at http://www.evrnu.com/#intro. (06 May 2016).

Federal Trade Commission. (2016). *Protecting America's consumers, wool products labeling Act* [Online]. Available at https://www.ftc.gov/node/119457. Accessed April 29, 2016.

Green Retail Decisions. (2012). *Recycled bottles turn into new line of Levi jeans* [Online]. Available at http://www.greenretaildecisions.com/news/2012/10/18/recycled-bottles-turn-into-new-line-of-levis-jeans. Accessed May 02, 2016.

Hanes for Good. (2016). *Environmental responsibility*. Available at http://hanesforgood.com/environmental-responsibility/. (16 May 2016).

Hawley, J. (2015). Economic impact of textile and clothing recycling. In *Sustainable fashion—What's next? A conversation about issues, practices and possibilities* (pp. 204–230). Bloomsbury Publishing Inc.

Hayes, L. L. (2011). Synthetic textile innovations: Polyester fiber-to-fiber recycling for the advancement of sustainability. *AATCC Review, 11*(4), 37–41.

Jimtex Yarns. (2014). *Jimtex yarns home page*. Available at http://www.jimtexyarns.com/our-yarns/#ecological. (13 May 2016).

Kincade, D. H., & Gibson, Fay Y. (2010). *Merchandising of fashion products*. Upper Saddle River, N.J.: Prentice Hall.

Lamicella, L. (2015). *Adidas to develop custom, recyclable sports apparel* [Online]. Available at https://sourcingjournalonline.com/adidas-to-develop-custom-recyclable-sports-apparel/. Accessed May 04, 2016.

Levi's®. (2012). Launching denim made of recycled plastic bottles. *The Textile Magazine, 54*(1), 70–71.

Martex Reclaimed Fiber. (2016). [Online]. Available at http://www.martexfiber.com/products/reclaimed-fiber/. Accessed April 24, 2016.

McGregor, L. (2015a). *Are closed loop textiles the future of fashion?* [Online]. Available at https://sourcingjournalonline.com/are-closed-loop-textiles-the-future-of-fashion/. Accessed May 03, 2016.

McGregor, L. (2015b). *Scientists successfully separate poly-cotton blend textiles* [Online]. Available at https://sourcingjournalonline.com/scientists-successfully-separate-poly-cotton-blend-textiles-lm/. Accessed April 27, 2016.

McGregor, L. (2015c). *Speedo closes the loop with swimwear made from fabric remnants* [Online]. Available at https://sourcingjournalonline.com/speedo-closes-the-loop-with-swimwear-made-from-fabric-remnants/. Accessed April 24, 2016.

McGregor, L. (2015d). *Timberland steps up recycled content in footwear; commits to sustainability* [Online]. Available at https://sourcingjournalonline.com/timberland-steps-recycled-content-footwear-commits-sustainability-lm/. Accessed May 17, 2016.

McGregor, L. (2016a). *H&M sustainability report stresses need for industry-wide collaboration* [Online]. Available at https://sourcingjournalonline.com/hm-sustainability-report-stresses-need-for-industry-wide-collaboration/. Accessed April 27, 2016.

McGregor, L. (2016b). *Nike raises its sustainability game, sets new supply chain goals for 2020* [Online]. Available at https://sourcingjournalonline.com/nike-raises-its-sustainability-game-sets-new-goals-for-2020/. Accessed May 03, 2016.

McGregor, L. (2016c). *Texworld USA: Consumer education is key to selling sustainable apparel* [Online]. Available at https://sourcingjournalonline.com/texworld-usa-consumer-education-is-key-to-selling-sustainable-apparel/. Accessed April 25, 2016.

Mitchell, M. (1936). *"Gone with the Wind"*. MacMillian Publishers.

Moore, L. (2011). Apparel makers think green. *Apparel Industry Magazine, 53*(6), 68.

National Cotton Council of America. (2016). *U.S. and world cotton economic outlook*. Economic Services—National Cotton Council. Available from http://www.cotton.org/econ/reports/outlook.cfm. (04 May 2016).

Nike. (2016a). *Environmental impact* [Online]. Available from http://about.nike.com/pages/environmental-impact. Accessed May 11, 2016.

Nike. (2016b). *Reuse-a-Shoe FAQS, sustainability performance* [Online]. Available from http://help-en-eu.nike.com/app/answers/detail/a_id/39600/p/3897. Accessed May 11, 2016.

Nike. (2016c). *Sustainability has become a game changer for Nike* [Online]. Available from http://about.nike.com/pages/our-ambition. Accessed May 11, 2016.

Patagonia. (2016a). *Environmental and social responsibility*. Available from http://www.patagonia.com/us/patagonia.go?assetid=110473. (28 April 2016).

Patagonia. (2016b). *Reclaimed cotton* [Online]. Available from http://www.patagonia.com/us/patagonia.go?assetid=102265. Accessed April 28, 2016.

Patagonia. (2016c). *Recycled polyester* [Online]. Available from http://www.patagonia.com/us/patagonia.go?assetid=2791. Accessed April 28, 2016.

Patagonia. (2016d). *Recycled wool* [Online]. Available from http://www.patagonia.com/us/patagonia.go?assetid=93863. Accessed April 28, 2016.

Payne, A. (2015). Open and closed-loop recycling of textile and apparel products. *Handbook of Life Cycle Assessment (LCA) of Textiles and Clothing*, 103–123.

Peters, A. (2016). *CoExist Levi's made the first ever 100 % recyled cotton jeans* [Online]. Available from http://www.fastcoexist.com/3059826/levis-made-the-first-ever-100-recycled-cotton-jeans. Accessed May 14, 2016.

Porter, N. (2016). *H&M Unveils the Swedish Olympic Team's New Uniforms* [Online]. Available at http://www.racked.com/2016/4/27/11517942/h-m-sweden-olympics-uniforms. Accessed July 21, 2016.

Repreve. (2016). *Repreve home page* [Online]. Available at http://repreve.com/brands. Accessed May 13, 2016.

Rivera, L. (2013). Anonymous green futures magazine. *A recycled bottle blend for jeans* [Online]. Available from https://www.forumforthefuture.org/greenfutures/articles/recycled-bottle-blend-jeans. Accessed May 03, 2016.

Shephard, H. (2016). *Sustainability now! Nike and Adidas hop on the green train.* Available at http://fashionunfiltered.com/news/2016/sustainability-now-nike-and-adidas-hop-on-the-green-train. Accessed May 13, 2016.

Sustainable Apparel Coalition. (2016). Member spotlight: Levi Strauss & Co [Online]. Available from http://apparelcoalition.org/member_spotlight/levis/. Accessed May 06, 2015.

Sustainable Brands. (2010). *Hanes launches environmental marketing, products, and website.* Available at http://www.sustainablebrands.com/news_and_views/articles/hanes-launches-environmental-marketing-products-and-website. Accessed May 10, 2016.

Tenjin. (2016). *Closed-loop recycling system ECO CIRCLE* [Online]. Available at http://www.teijin.com/solutions/ecocircle/. Accessed April 24, 2016.

Vadicherla, T., & Saravanan, D. (2014). Textiles and apparel development using recycled and reclaimed fibers. In *Roadmap to sustainable textiles and clothing* (pp. 139–160). Springer.

Velasquez, A. (2014a). *Adidas reveals prototype shoe made of recycled ocean waste* [Online]. Available at http://vampfootwear.com/adidas-reveals-protoype-shoe-made-of-recycled-ocean-waste/. Accessed April 30, 2016.

Velasquez, A. (2014b). *Cone Denim and Unifi launch eco-friendly, soft stretch Denim* [Online]. Available at https://sourcingjournalonline.com/cone-denim-unifi-launch-eco-friendly-soft-texstretch-denim-av/. Accessed April 20, 2016.

Yin, Y., Yao, D., Wang, C. & Wang, Y. (2013). Removal of spandex from nylon/spandex blended fabrics by selective polymer degradation. *Textile Research Journal.*

Zamani, B. (2011). Carbon footprint and energy use of textile recycling techniques, M.S. Thesis, Chalmers University of Technology.

Denim Recycling

Shanthi Radhakrishnan

Abstract The magic of denim jeans has overpowered the global right from the day it was invented. History has seen its development through many ages and periods. Sustainability and recycling is the buzz word today, and all stake holders in the apparel supply chain right from manufacturers to consumers are working toward this cause. This focus has called for research and development all over the world to undertake many issues related to denim recycling to make the best use of used materials for new product development. Many retailers take immense efforts to showcase their involvement in the closed-loop recycling initiative by encouraging consumers to bring back old used garments for new ones and converting these garments for the manufacture of raw materials or intermediate substances. Manufacturing compostable jeans without the use of nylon threads and rivets shows the change in the manufacturing process, and new technologies are in the pipeline to recycle fibers from denim with unchanged quality. The governmental support has also been extended by means of many programs on solid waste management, reduction of load to landfill by reuse and recycling and laws and regulations for environmental protection. This chapter deals with the importance of denim and its impact on society, manufacturing and landfill issues, the technologies involved in the reuse and recycling of denim, the appraisal of work done by many organizations around the globe to recycle denim for regenerated textiles and reclaimed products along with the roadmap for denim recycling in terms of sustainability. Denim, one of the most widely used material in the world, has significant impact on environment in manufacturing and waste management stage. Denim recycling has opened vast opportunities for savings in the use of raw materials, energy and water consumption, chemicals and auxiliaries and waste water treatment. Reprocessed fibers from denim waste have the coloration from the raw material used, and hence, dyeing and finishing processes can be eliminated to a great extent. Many leading retailers like H & M, Adidas and Nike showcase and market their products with the percentage of recycled material in the product profile; their statistical reports reveal the quantity of clothes they have collected from their shoppers and the amount they

S. Radhakrishnan (✉)
Department of Fashion Technology, Kumaraguru College of Technology, Coimbatore, India
e-mail: shanradkri@gmail.com

© Springer Science+Business Media Singapore 2017   79
S.S. Muthu (ed.), *Textiles and Clothing Sustainability*,
Textile Science and Clothing Technology,
DOI 10.1007/978-981-10-2146-6_3

contribute to international charity from their proceeds. Raising the awareness of the consumer's contribution toward sustainability and environment safety has paved the increase in recycling of not only denim but also many other materials that are used today. The focus should move toward manufacturing products without waste, and if waste occurs, it should be recycled to lead to a zero waste economy.

Keywords Sustainability · Denim reuse and recycling · Recycling economics · Zero waste management

1 The Denim Background

1.1 History of Blue Jeans

A strong fabric, denim, with innumerable variants beyond one's imagination, is the most popular textile of all ages. It can roll out as any apparel stretching from work wear in factories to high fashion designer clothing in catwalks. The transformation of denim to jeans was due to the joint efforts of two men Jacob W. Davis, a tailor, and Levi Strauss, an entrepreneur. From then onward jeans transformed many times over the years in terms of shape and cut, meaning and symbolism. The American Fabrics magazine in 1969 stated that denim was one of the oldest fabrics yet proved to be 'eternally young' due to continuous use and interest (Downey 2014).

Originally blue jean material denim was a strong material made of wool. In 1700, it was blended with cotton and later on it was made as 100 % cotton material. The name was derived from serge de Nimes which meant serge of Nimes, France. This material was used as sail cloth in ships and thanks to the innovative thoughts of some Genovese sailors who converted into 'genes' which later became jeans. Indigo was a blue dye obtained from the indigo plant and was used as a dye from early times—2500 B.C. This dye was imported from India till the twentieth century after which the synthetic dyes replaced the natural dyes. In earlier times denim and jeans were used to identify fabrics. Denim has a dyed warp yarn and a white weft, while jean had both warp and weft yarns dyed.

1.2 Interesting Facts About Blue Jeans

The oldest pair of jeans was found in 1997 aged 100 years. Two styles of jeans were presented to the market as indigo blue and brown cotton 'duck,' and the first jeans were called 'waist overalls.' Indigo color was chosen to hide the dirt on jeans. In total, 20,000 tons of indigo dye was produced annually to color jeans though only a few grams of indigo dye is sufficient to color one pair of jeans. A bale of cotton can produce up to 225 pair of jeans. The first label to be attached to a garment was a red flag attached to the back pocket of Levi Strauss jeans. A retailer named 'Limbo' in New York East Village washed jeans to get the worn-out look, which became a great hit in the fashion field.

About 37 sewing operations are necessary to construct Levis 501 model of jeans. Zippers were placed in the front of jeans for men and at the sides in jeans for women. Orange thread was trademarked and used for stitching of jeans manufactured by Levi Strauss and Company to enhance the copper rivets. Rivets were originally fitted at the corners of all pockets to make them stronger. Complaints came in that rivets scratched saddles and chairs, and hence, they were removed at the back of jeans. The patent for blue jeans was obtained by Jacob Davis and Levis Strauss and company on May 20, 1873, to be considered as the birthday for blue jeans (Anonymous 2016a, b). On an average every American owns 7 pairs of blue jeans. Approximately 450 million pairs of jeans are sold in the USA every year, and over 50 % are manufactured from Asian countries like China, India and Bangladesh. Uses of synthetics as dyes and as elastane for skinny jeans have stepped into jean manufacturing. The popularity of jeans spread far and wide outside the USA when the American Soldiers in World War II wore jeans in their off duty times. The most expensive pair of jeans was sold for $250,000, and the longest pair of blue jeans was 68 m long (Anonymous 2016a, b).

1.3 Soft Value of Denim Jeans

Any clothing brings emotions when worn, e.g., school uniform, soldier's uniform, tailored suit or a pair of jeans. Feeling of 'comfort,' simplicity and informality seem to depict the attitude of many while wearing jeans. This may be due to the unification of cloth and human form, and the wearers continue to feel their own body image. These feelings are highly valued in the contemporary western society where self-identity and social receptivity are important. Further as time passes, jeans become more human and personalized with the individual.

The significance of the color blue in Western society is innumerable. The symbolic association with history includes the woad and the barbarian, the Virgin Mary, the sea and the sky and the color of our planet (Candy 2005). Denim blue trousers, a Victorian style of clothing, has surpassed all garments made of blue and gained widespread cultural connotation over the decades.

Denim jeans can throw up a wide range of emotions as it is very close to the personal side of many people both young and old. A study conducted by Professor Karen Pine showed that there is a strong link between the mood state and clothing; clothing is used to reflect the mood of a person. People who do not wish to stand out in a crowd or when depressed wear jeans that may be poor in cut and ill-fitting. The research highlights the correlation between depression and wearing jeans (Cooper 2016).

1.4   Social and Cultural Value of Denim Jeans

Denim jeans were originally workwear pants and not opted by the rich and affluent class. In the early 1920–1930, blue denim jeans were used as an index of difference against the white peasant *calzon* worn by the working class in Mexican villages (Comstock 2016). The sequence of events that followed in America transformed blue jeans into an icon of depression of American Democracy. During World War II the US military consumption of jeans and the improvised industrial regulations aided in an increase in economy by the manufacture of denim jeans.

In America during the golden era of cinema, great actors like James Dean and Marlon Brando wore jeans because they wanted to wear them and not as a symbol of workwear (Little 2007; Hegarty 2012). A series of cultural changes (magazines, movies and rock 'n' roll music) shook the entertainment industry, and denim jeans became a mode of expression. Denim jeans promoted by Hollywood stars became a sex symbol and tight jeans, a symbol of youth in revolt.

By 1960s jeans were widely accepted and became less stiffer and softer than the jeans of 1950s representing softer values (use of softer and lighter colors) and moving from artificial to natural look (use of browns, soft blues and greens). During the 1970s music and pop culture gave rise to bell-bottom jeans; it became a status symbol with the introduction of designer jeans and the prices were as high as $100 (Sullivan 2006; Weber 2006). Denim jeans succeeded throughout the ages as it was convertible, adaptable and malleable to suit the different needs of the ever changing cultures. It became a powerful tool in America during the civil rights era to resist the old American traditions and unified different types of people (both male and female, rich and poor, black and white) onto a common ground—wearing blue denim jeans (Blue Jeans 2011).

In Sweden denim brands are closely linked with the cultural values of the people terming this phenomenon as 'The Swedish denim miracle.' Denim Demon, one of the latest additions to the band wagon, bears names from the Sami language as a tribute to the natives of northern Sweden, Norway, Finland and Russia. 'Nudie' is jeans worn as second skin. The philosophy highlights passion and deep relationships, and as time passes, it is supposed to be shaped and become one with the wearer.

1.5   Individualistic Intimacy of Blue Jeans

Blue jeans are very close to the individual, and the bond strengthens as time passes. A Chinese survey revealed that one-third of the 219 respondents felt high popularity existed for selection of jeans as causal wear and fit and comfort were important to an individual while selecting jeans (Wu and Delong 2006). Sensory pleasures like vision and touch inspire and stimulate the consumer to try on and feel the product before purchase. Youngsters look out for a distinctive design to enhance their social significance, while color provokes emotional responses and aesthetic appeal (Schmitt and Simon 1997). A survey taken among Canadian consumers reveals that 96.1 % (n = 365) favored denim jeans and over 50 % wore it on a daily basis, and on an average each respondent had 9–10 pairs of jeans (Rahman 2011). A study conducted in Ahmedabad, India, highlighted the preference of wearing jeans four to five times a week by 60 % of the respondents, while 27 % wore it on daily basis. Jeans were also the preferred garment by the married Indian women (Upadyayay and Ambavale 2013). The other features which attracted the female respondents to denim were freedom of dupatta (a long scarf round the neck), companionship with the jeans during happy and sad times, comfortable at workplace, tight fit jeans gave a feeling of wearing a churidar (a tight fitting Indian garment for covering waist to ankle), perfect style for casual wear and the preferred color was blue. This study highlighted the deeper meaning linked with jeans in the minds of the respondents.

Fit is another aspect of jeans which determines its success or failure. Factors which contribute to a sense of good feeling while wearing jeans are ease of garment, figure flaw reward, sex appeal, body cathexis and sensory pleasure. Body image and self-esteem are closely related to body cathexis. The mental picture of one's own body at a moment in time is body image, while self-esteem is the way we feel about ourself. Body cathexis is the degree of satisfaction or dissatisfaction one has toward one's body parts and the whole body (Robinson 2003).

Jeans are an important tool to evaluate one's body image and self-esteem to result in body cathexis, and when this turns out to be good and pleasant, the product is a big hit. Dissatisfaction usually arises from the lower torso comprising of the hips, thighs and buttocks. The fit that you want will determine the body image and self-esteem. Slim fit, skin fit, regular fit, comfort fit, semi-baggy fit and baggy fit are some of the fit classes available to the consumer. Levi's has also introduced three variations in cut and fit based on the curve that define the hip to waist—Curve ID collection. The shape of the customer is defined to select the custom fit (Levis 2016). The jeans are contoured to follow the natural shape of the hip and ensure the fit at the waist avoids pinching or gaping. Denim jean companies offer a wide range of styles, fit and size assortments to cater to the different needs and contours of today's customers. Low cut jeans with figure hugging fit enhance the appeal of a person highlighting the curves of the body to create an impact on the opposite sex. The ruggedness of the male and the femininity of the female have been used to enhance the appeal of jeans by many manufacturers and retailers. Calvin Klein decided to combine sex appeal and youth in their advertisement 'nothing comes between me and my Calvin' to create a sensation among the youth and thereby promote sales.

Touching a product activates consumer attitudes and behavior. Sensory feedback of touch tends to increase confidence in evaluation of the product and persuades the purchase intentions toward the product. The placement of products in a store is closely linked with touch. Online virtual stores do not have this advantage while selling products. Attitude toward behavioral acts may be utilitarian or hedonic in nature. Utilitarian aspect is an attitude related to usefulness, value or wiseness of the behavior as perceived by the consumer; hedonic attitude develops from the pleasure experienced or anticipated from the behavior apparent to the consumer. More than utilitarian, the hedonic benefits to customers are seen in sales promotion and retailing and the use of touch as a hedonic tool is applied in a wide variety of products and services (Ahtola 1985; Peck and Wiggins 2006). A pair of jeans is valued for the finishes it undergo to add value to its sensory appeal and softness. Hedonic aspects of jeans are extremely important as the manufacturer spends a large amount of resources to capture the required final appearance. Finishes like enzyme wash, fraying garment dyed, hand sanding, silicone softener finish, sand blast, vintage look, stone wash, tint, torn or worn-out look, whiskering and nanofinishing add to the tactile feeling of the jeans (Denim future 2016; Belly 2011). In some cases a combination of treatments is given to acquire certain feels of the fabric. The elements of touch and feel provide unexpected information on the experiential and aesthetic components of the jean and help if sealing the gap between the individual and the product.

The closeness of the jean to the body makes it an integral part to picturize the emotions, balance, feelings, kinship, sensory appeal and self-esteem of an individual among his peers and in society. The revealing yet concealing nature of jeans can entice an individual to develop an attachment toward it to create a magical bonding that has left a lasting impression on the individual to make it a classic of all times.

1.6   Life Cycle of Denim Jeans

- Cotton is the raw material for jean production harvested in around 90 countries around the globe. It requires 20,000 L of water to grow one kg of cotton, and a lot of pesticides are used in cotton cultivation. Unsustainable farming practices and heavy usage of water have polluted the eco-systems and environment where cotton is cultivated. The life cycle of jeans is represented in Fig. 1.

Fig. 1 Life cycle of jeans (Op 2014)

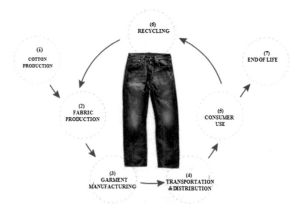

Table 1 Environmental impact of 501 Levi's jeans (Levi Strauss & Co. 2015)

Sl. No.	Environmental parameters	Quantity	Equivalent
1.	Carbon dioxide equivalent for measuring carbon footprint	33.4 kg CO_2-e	Sixty-nine-mile drive in an average US car/246 h of TV on a big plasma screen
2.	Consumption of water	3781 L	Three-day water requirement of the total households in the USA
3.	Eutrophication	48.9 g PO_4-e	Total amount of phosphorous in 1700 tomatoes
4.	Land occupation	12 m^2 per year	A square with one side equal to length of 7 people with outstretched arms with fingertips touching each other

- Cotton from the fields is taken to the textile mill where the fiber is converted to yarn by the ginning, carding and spinning processes and then to fabric using one white yarn and one colored yarn usually dyed in blue color. Sometimes synthetic material like polyester or spandex may be included to provide for ease of care or stretch. The twill weave is used to weave denim fabric.
- The fabric is cut into different components and assembled into jeans in apparel industries situated in different countries like Bangladesh, India, China and many others. Sewing is done through assembly lines or through the unit production system where sewing operators work component by component that the finished garment is out at the end of the line.
- The finished jeans are checked for quality, packed and transported to different parts of the world and distributed through wholesale, retail and online centers.
- The use phase by the consumer requires care and maintenance of the jeans. The impact on the environment can be reduced by line drying and cold water wash.
- The recycling of jeans is a recent concept adopted by consumers, retailers and manufacturers to make use of jeans after its use with the consumer.
- When recycling is not possible, then the jeans are sent to landfills. It has been estimated that 22.8 billion pounds of clothing ends up in landfills every year.

The entire life cycle impact of one pair of 501 Levi Strauss denim jeans is given in Table 1.

The sustainable practices followed by Levi's are designed to reduce the environmental impact of manufacturing jeans. The climatic changes like drought and storms have brought about failure of cotton crop and spiraling prices. Manufacturers are looking out for sustainable raw material. BCI cotton and fiber from recycled PET bottles are used as sustainable raw materials. BCI cotton uses 70 % less pesticides than conventional cotton (Kobayashi 2013). Intensive water recycling has been taken up by Levi's, and about 75–80 % of the water is recycled by microfiltration and has developed a standard for its supplier companies. Eco-friendly practices are also followed in of dyeing and finishing where enzymes are used instead of conventional chemicals.

Food. Water. Denim. Let's get back to the Essentials.—Anonymous
Reference: https://s-media-cacheak0.pinimg.com/236x/ce/09/71/ce097165a3b6a4b0bb2a0e 621b200945.jpg252

Jeans represent democracy in fashion—Giorgio Armani
Reference: http://izquotes.com/quotes-pictures/quote-jeans-represent-democracy-in-fashion-giorgio-armani-6934.jpg

These words highlight the importance and impact of jeans in the lives, visions, aims and achievements of all throughout the world. The journey of blue jeans which began as rough clothing workwear was nurtured to great heights by the servicemen and heroes of cinema, and this wave boomeranged by the rebellious youth who expressed their attitude and defiance through this magical classic which has survived all the years though many fashions came and went without much ado. The denim fabric wore well, was comfortable and reasonably affordable to all classes of society. Over the years millions and trillions of jeans have been manufactured and used, but today the consumer has turned his attention to the recycling of jeans to extend its life and to save the environment from further damage.

2 Problems with Textile Waste—Denim Waste and Associated Issues

Waste is any material which is of no use to the organization it belongs and may end up in being thrown away, thereby creating problems to other segments of the society and environment. Waste for one sector may be a prospective reusable resource for another segment. The matter of great concern today to most industries, community and governments is the magnitude of waste that is generated and its negative impact on environment. The state-wide solid waste composition and characterization study carried out in 2009 has reported that textiles and clothing contribute to approximately 96,500 tons (4 %) of the waste stream in Connecticut, USA; further 71,800 tons (74 %) of the textile waste is from residential sources and 24,700 tons (26 %) originates from non-residential sources. It has also been estimated that $5.7 million is being spent as waste disposal tipping fee at the rate of $60 per ton in Connecticut (DEEP 2016). Why is any textile waste a problem? When this question is considered, a multitude of reasons is accompaniments of textile waste, namely the harm to human and environment by waste disposal; lack of space in landfills for any waste including those generated from textiles and apparel; the cost and depletion of valuable resources which are slowly becoming scarce; the moral view associated with waste generation and over consumption and so on.

2.1 Types of Waste

The Environmental Agency has classified waste as controlled and non-controlled waste. Waste generated from households, commercial and industrial organizations and from construction come under the category of controlled waste, while waste resulting from agriculture, quarrying, dredging and mining belongs to the non-controlled waste. It has been reported that between 1993 and 1996, the mean production of daily controlled waste in the European Union member states was around 370 kg/capita/ annum (Fischer 2000). Waste has been classified on the basis of generation as pre-consumer textile waste, post-consumer textile waste and industrial textile waste. Pre-consumer waste is the remaining production processes in the industry which includes raw material to finished products ready for market. This includes offcuts, shearing, selvedges, rejected materials, b-grade garments, export surplus which are homogenous and clean in nature to be used for other purposes. They may be sold at low rates to agents who deal with these materials or may be sent to landfills. The waste under this class has great potential for reuse and recycling. Figure 2 shows the types of waste and the reuse and upcycling methods.

The post-consumer textile waste can include any product that has completed its life cycle and is no longer useful to the consumer in both function and aesthetics. The consumer no longer needs the product and decides to discard it due to wear and tear, damage, ill fit and out of fashion. In most cases the garment or product that is reasonably good can be recovered and used as second-hand clothing and sold to the third world countries which host huge markets for second-hand clothing. Unused products are shredded and converted to raw material for further use in manufacturing.

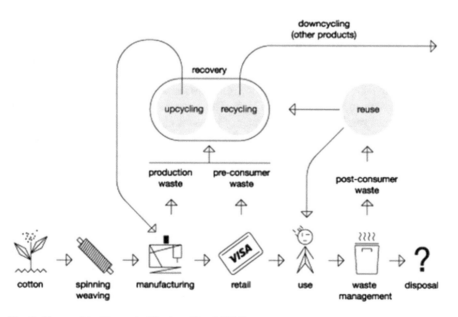

Fig. 2 Types of textile waste (Trash to Trend 2016)

Fig. 3 Sludge waste for disposal in textile dyeing industry (KNITcMA 2011)

Industrial textile waste is the results of the manufacturing processes and is termed as 'dirty waste.' This waste has collection and contamination issues and may not be recovered. An example that has created immense problem to the health and environment is the mounds of solid waste that has been piled up in textile dyeing industries ready to be disposed to landfills as shown in Fig. 3. However, all agencies concerned are involved in activities to avoid this mass of waste and opens up huge opportunities for research (Caufield 2009).

2.2 Denim Waste and Associated Issues

Denim jeans are usually made of cotton, and this leads to the understanding that it will deteriorate in the environment. Cotton has properties which help in deterioration of the raw material as it is easily destroyed by mold, acids, bleach and fire. These scientific facts are true, but in practice a tough pair of jeans made with 100 % cotton fiber can stay alive for a long time in the environment and the negative environmental impression is very high. The estimate is that 450 million pairs of jeans are sold yearly in the USA; a pair of jeans requires 150 lbs of cotton and 1500 gallons of water (Gordon and Hsieh 2007). Apart from this a large quantity of resources are used for manufacturing of raw material into end product combined with the chemicals used for the special denim finishes sand blasting, aging and distressed looks. Denim jeans consumption has spread throughout the world, and the requisites are very high. Further hazardous chemicals like cadmium, mercury and lead, the remnants of denim finishing, are dumped into water resources. China and India, the highest producers of textiles, are facing severe consequences of unmanned manufacturing which has resulted in polluted water and land resources; clean drinking water is a rarity for the locals in these manufacturing centers.

Denim with a worn-out look is termed as distressed denim. The process involved coloring with toxic dyes, acid baths and sand blasting followed by a bath with chemicals. Depending on the degree of worn-out look required multiple treatments are given to the jeans which use lot of water and chemicals, and the resultant effluent is harmful to the environment and society. Tehuacan, the heartland of Mexico's denim industry, houses many industrial laundries that add the finishing touches to jeans. Bleaches, dyes and detergents are commonly used agents for laundering and are considered to be highly polluting (Creek Life 2016). Unregulated management of these industries has turned the Mexican rivers to become literally blue and has a negative impact on the environment. When consumers become aware of the degrading elements of denim production, they will be motivated to choose more eco-friendly alternatives.

2.3   Indigo Dyes

Synthetic indigo dyes attach themselves to the threads of denim yarns externally with the help of a mordant. Mordant may be of various types, but metal mordants like chromium or aluminum are commonly used. The waste water, with these mordant remains, from the denim industries can kill plants, destroy ecosystems and poison drinking water. Of the two mordants aluminum is better than chromium in terms of environmental pollution. These mordants are responsible for the fading characteristic quality of a pair of jeans during the wear cycle.

The natural indigo dye is obtained mainly from the plant *Indigofera*, a tropical plant. The *I. tinctoria* was the species domesticated in India. The plant leaves, which contain about 0.2–0.8 % of the dye compound, are fermented to convert the glycoside indicant to a blue dye indigotin. The precipitate from the fermented leaf solution is mixed with lye, pressed into cakes, dried and powdered. The powder is mixed with other substances to produce shades from blue to purple. Indigo is a challenging dye as it is insoluble in water. The reduction process helps to dissolve the dye, and it is converted to 'white indigo' (leuco indigo). The oxidation process helps in the coloration of the indigo blue. The fabric removed from the white indigo dye solution combines with the oxygen in the air and converts to the insoluble, intensely colored indigo blue (Schorlemmer 1874).

Indigo was termed a blue gold, and in the nineteenth century 7000 km^2 had indigo cultivation mainly in India. Indigo plantations supported slavery, and during the British period in India, the peasants in Bengal revolted against unfair treatment which gave rise to the Indigo Revolt in 1859 (Kriger and Connah 2006; Steingruber 2004). Natural indigo, with this historical name and fame, is not an environment-friendly dye. It takes a very long time to decompose and darkens the river water blocking the entry of sunlight and oxygen to the fauna and flora of the water body. In 1897, 19,000 tons of natural indigo was produced from plant sources. The love for the blue has brought about heavy usage of natural indigo till the synthetic indigo was commercially manufactured by BASF in 1897.

2.4 Dyeing Pollution—Savar, Bangladesh

Bangladesh is the second leading clothing exporter in the world after China, and many leading retailers like Walmart, J.C. Penny and H & M have a strong base of suppliers from Bangladesh in their supply chain. The low cost formula includes paying the lowest wages in the world, spending minimal on work conditions and safety, allowing untreated effluent discharge to save utility costs and not following the environmental regulations. This attitude has taken its toll on the environment as the water pollution disaster in and around the capital Dhaka. Interlacement of toxic waste water with the paddy fields, dying fish stock, garbage filled waterways, pollution in the food chain are some of the effects of the full discharge of untreated water into the water bodies. Fines have been levied against the textile and dyeing industries, but no serious action has been taken by the governmental agencies. Factories in these areas continue their operations without an environmental clearance certificate for more than 23 years. The pictures in Fig. 4 show the heavy damage to environment in Savar, Bangladesh. The toxic stench and the odor from the canal affect the people living in the area as the air is choking with chemicals, and the school students in those areas suffer from light headedness and dizziness. The latest trend in color is well understood by the school students by just looking at the canal (Hasan 2013). Greenpeace had released an overwhelming report titled 'Toxic threads: The Big Fashion Stitch-up' which highlighted the hazards of dye pollution and jeans, and the top offenders of toxic dye use were noted as Levi's and

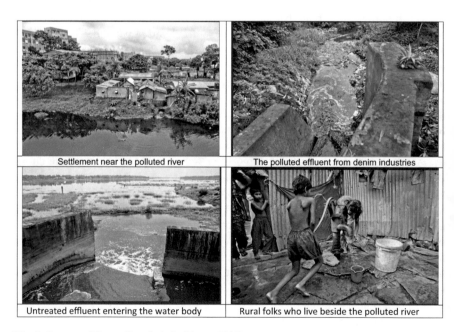

Settlement near the polluted river | The polluted effluent from denim industries

Untreated effluent entering the water body | Rural folks who live beside the polluted river

Fig. 4 Images of Savar, Bangladesh (Hasan 2013)

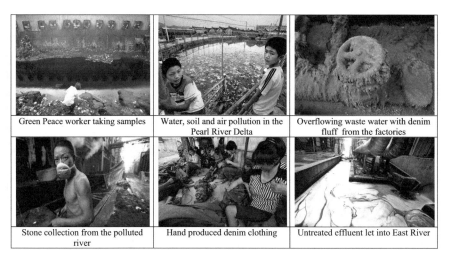

Green Peace worker taking samples	Water, soil and air pollution in the Pearl River Delta	Overflowing waste water with denim fluff from the factories
Stone collection from the polluted river	Hand produced denim clothing	Untreated effluent let into East River

Fig. 5 Images of pollution from denim production, China (Guang et al. 2013)

Tommy Hilfiger (Barker 2013). All the reports on Bangladesh topping the world's jean production are interlinked with sad tales of environmental damage to the country, people and environment (Fig. 5).

2.5 Pollution in Xintang, South West China

Xintang, South west China, manufactures one-third of the total jeans sold in the world and two third of denim clothing produced in China. This town is titled 'the denim capital of the world.' The factories in this town produce 300 million denim articles a year and employ 220,000 people (Guang et al. 2013). The waters of the East river in Xinyang have turned blue with a stinking smell due to effluent from bleaching, dyeing and washing factories. The waste water treatment plant has not been functioning for over a year. Many denim manufacturing industries have been shifted to the Xinzhou Environmental Industry Park which has environmental standards far below the stipulated level. Xinzhou Village which was fertile and well known as a fishing village has no sign of farming. The environmental park in Xinzhou Village has noise pollution, ditches with dark blue filled water, dust on roads as light blue and with black river waters. People are offered 2000–5000 yuan ($325–815) per month, but it is difficult to get labor due to high pollution. The air has the potassium permanganate smell used for spraying on the jeans, for worn-out look is inherent with the workers and the air they breathe has denim dust. The laborers have neither contracts nor insurances, and they receive their first wages after three months. Workers and people living in these cities face problems like infertility, skin and lung problems. The East river is a major source of drinking water and for millions of people, and the water securities of cities downtown the river like Dongguan and Shenzhen are also threatened by pollution associated with denim manufacture.

2.6 Problems Associated with Waste Management

Waste management consists of collection, processing, transport and disposal of generated waste. Methods like recycling, composting, sewage treatment, incineration and landfill have been used consistently depending on the type of waste generated and the extent of recycling possibilities. World Health Organization (WHO) suggests that substances can be termed as pollutants based on their noxiousness, environmental persistence, movement, bioaccumulation and other dangers like explosiveness (WHO 2000). Health issues are a main problem for the workforce involved in waste management and those civilians who live near the waste dumping sites. Exposure levels and duration of exposure to substances like cadmium, arsenic, chromium, dioxins, nickel and polycyclic aromatic hydrocarbons (PAH), not only produce carcinogenic effects but also have serious implications on the central nervous system, liver, kidney, heart, lungs, skin and reproduction of the people near the waste sites. Further pollutants like SO_2 and coarse dust particles (PM_{10}) cause disease and death to vulnerable groups of the society like infant and the elderly, while dioxins and organochlorides are lipid-loving molecules which tend to get settled in fat-rich tissues resulting in serious reproductive or endocrine disorders (Rushton 2003).

Epidemiological studies have shown the ill effects of landfill sites. During 1930s and 1940s large quantities of waste were dumped at Love Canal, New York, which included toxic waste from pesticide production. By mid-1970s, leaching of chemicals was detected in streams, soil, sewers and air quality in houses that were built around the landfill site in 1950s leading to research and reports of such areas. Apart from problems like nuisance, odor, risk and stress, birth defects, reproductive disorders, congenital malformations and cancer are adverse effects of landfill options. Incidence of low birth weight babies and neonatal deaths was found in the population around the Love Canal site and also in the Lipari landfill site in New Jersey due to heavy dumping and the resultant exposure (Vianna and Polan 1984; Goldman et al. 1985; Berry and Bove 1997). In UK, the Welsh landfill of Nant-y-Gwyddon health issues increases ninefold and included neural tube defects, chromosomal disorders, malfunctions of the cardiac septa and anomalies in the arteries and veins (Dolk et al. 1998; Vrijheid et al. 2002). The third largest site for waste dumping Miron Quarry site in North America showed the incidence of cancers in liver, kidney, pancreas and is connected to the closeness of the inhabitants to the site (Goldberg et al. 1995, 1999). Further the work force in the waste sites faces health risks related to gastric and lung cancer and skin problems due to VOCs, bioaerosols and dust levels (Gustavsson 1989; Rapiti et al. 1997).

Pollutants associated with incineration of waste are particles, acidic gases, nitrogen dioxide (NO_2), sulfur dioxide (SO_2), aerosols, metals and organic compounds. Health risks due to incineration of waste are due to the pollution pumped into the air resulting in cardiovascular and respiratory illness, injury, disease and death depending on the duration of exposure. Bronchitis, reduction in functions of the lungs, respiratory problems, shortened span of life and lung cancers are respiratory disorders coupled with incineration of waste (Dockery and Pope 1994; Katsouyanni et al. 1997). Organic compounds like dioxins and polychlorinated biphenyl (PCBs) are highly risky since they have the tendency to accumulate in the body leading to skin diseases like leukemia and chloracne, cardiovascular diseases and laryngeal and childhood cancers (Elliot et al. 1992, 1996, 2000). Studies have shown that high levels of dust particles have resulted in deaths with standardized mortality ratios (3.35) at 95 % confidence level 1.62–1.64 in Sweden incineration sites due to lung cancer, while deaths with standardized mortality ratios (2.79) at 90 % confidence level 0.94–6.35 were reported at the Italian incineration sites due to gastric cancer (Gustavsson 1989; Ancona et al. 2015). However, epidemiology studies are using the biomarker technology for estimating the level of exposure and analyzing the biological response or effect due to the respective exposure level. Alternative technologies in thermal waste treatment such as gasification and pyrolysis and biomechanical pretreatment of waste before disposal are used widely and give scope to great research opportunities to understand the effects and results of adopting these technologies.

Industries manufacture products for consumers, but at what cost? Sustainable production can be termed as a way of business that keeps what we are currently enjoying from nature for our future generations. Many instances in history have revealed the effects of unethical manufacturing and the impact on environment which in turn affects all living organisms and humans. Most people have concern for the environment, but they feel that someone else will save our world. Our future cannot go up in smoke, polluted smoke. All stake holders concerned should follow the concept of being part of the solution and not part of pollution. As the levels of production increase to showcase, the strongest industrial nation so too is the pollution problem. When environmental problems are foreseen and taken care of and organizations are ready for any unpredicted dangers, then all industrial activities will flow with a direction toward sustainability which is the focus of today's world.

3 Reuse and Recycling of Denim—Technology of Denim Recycling

The main constituent of all environmental and economic doctrines today throughout the globe is sustainability and development. Many problems in the environmental sector are due to the increase in production and consumption of products coupled with material movements. The use of natural resources in both renewable and

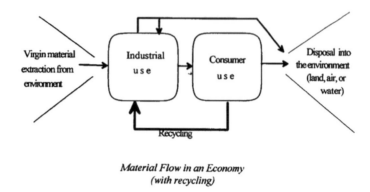

Material Flow in an Economy
(with recycling)

Fig. 6 Material flow in an economy with recycling (Shore 1994)

non-renewable form as raw materials, energy, water and land is related to the production and supply of goods which is in turn accelerated by population growth and urbanization. This cyclic process of high population growth and consumption exerts high pressure in the developed world resulting in an increase in environmental issues.

Prevention of waste creation and recycling of waste provide maximum benefits of utilization of products by extending their usage and life before being discarded as waste. Material flow in an economy is termed as unidirectional when products are manufactured used and thrown away as waste into the environment without any concerns. When the recycling component is included, it helps to absorb the residuals of industrial and consumer use as shown in Fig. 6 (Shore 1994). Recycling programs are important due to economic, social and environmental reasons. Waste disposal programs are more expensive when compared to recycling packages. The cost of manufacturing products has become very high due to scarcity of natural resources making recycling more rewarding than new product production.

The social goodness of recycling is creation of jobs to the lower economic groups. It has been estimated that recycling creates four jobs for a single job in the waste disposal sector. It also helps small and family enterprises and creates community bonding with people working toward a common cause—environmental issues. The Environmental Protection Agency (EPA) has reported that in the USA, the reuse and recycling industry has employed 1.1 million people with an annual payroll of nearly $37 billion, earning $236 billion as annual revenues. Further recycling also encouraged support industries providing 1.4 million jobs with $52 billion as payroll to produce $173 billion as receipts (Beck 2001). Recycling saves energy, water and resources leading to sustainability by reducing greenhouse gas production, fossil fuel usage, destruction of natural surrounding conflicts related to land use (Cuc and Vidovic 2011; WWF 2008).

3.1 Source of Denim Waste

The cutting waste during denim jeans production is unraveled and reused as raw material for the production of weft yarns in denim fabric manufacture. The dark indigo blue of the waste is retained as the waste is not subjected to washing or finishing processes involved in jean production. The strength of these yarns may be reduced marginally due to the mechanical processes involved in recycling process.

Another source of denim waste is the post-consumer denim jeans where the heterogeneity of the material makes recycling complex in nature. Color, quality of fabric and garment accessories like rivets, buttons, zippers and labels are the key components which add to the heterogeneous nature of denim waste. In most instances, post-consumer denim jeans are salvaged as clothing applications. Problems also arise in the collection and sorting process due to worn-out labels, and most of the recycled denim jeans are shredded and used for the manufacture of low-value products like thermal and sound insulation and pressure distribution. Growing interest among consumers, manufacturers and retailers has led to the usage of recycled materials for sustainable products due to stringent policies in the global market. Many research organizations like Texperium (NL) (Paul 2015) have turned their attention to value-added recycling of post-consumer denim jeans for attaining the goals of resource efficiency and green procurement.

3.2 Repurpose and Reuse Denim Jeans

The most important need is to expand the use phase of a garment. The popular concept for the closed-loop mind set is reuse, repair, recycle, redesign and reimagine as shown in Fig. 7. Giving clothes to a second-hand store or adding a vintage label for extending the use of a product forms an integral part of reuse.

Fig. 7 Closed-loop concept

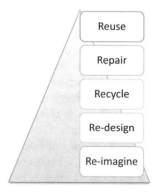

Reuse and repair are interlinked as repair can land up in reuse or further use. Recycling adds a new phase to a product, a product which was no longer in use becomes useful. It helps in supplying raw material for production, and a combination of virgin material with the recycled material would help in maintaining the quality of the product. Two important aspects of redesign are 'Design for Recycling (D4R)' and 'Recycling in Design (RiD).' In D4R, the bottlenecks for recycling are removed by changing the design lightly to solve the problems, e.g., detachable buttons may be used in jeans like cuff links; material composition and care instructions can be printed inside of pockets for long-lasting effects. An example for RiD is the use of recycled fibers in jeans for G Star (Paul 2015). Glue jeans offer seams that are glued instead of being stitched in a LSa seam. This facilitates mechanical recycling without many hindrances.

3.3 Upcycling of Denim Jeans

The term upcycling is given to the transformation of byproducts, unwanted products or waste materials into new materials or products of better quality or better environmental value. When products are broken down or dismantled, their value is lost and the materials may loose their quality in some manner due to the process followed. They are graded for a secondary or tertiary value when compared to their virgin counterparts. This is generally known as Recycling–down cycling. This is a regenerative process where the end products are cleaner and can be blended with other materials for new product development. A good example of down cycling is the production of polyester fibers from PET bottles, but the materials obtained by this process have structural weakness like strength loss and lower melting point than virgin polyester fibers. However technological improvements have come to the aid of the plastic industry and processes like conversion of waste plastics into paramagnetic conducting microspheres, into carbon nanoparticles by applying high temperature and chemical vapor deposition or treating polymers with electron beam doses of 150 kGy to increase the bending strength and elasticity of the plastics (Mondal et al. 2013; Altalhi et al. 2013) (Fig. 8).

Many products can be made by creative designers with aesthetic and technical prowess. Handwoven cotton and denim rugs, table runners, accent rug or door mat, can be woven into washable machine-dried products. Innumerable items like hand-quilted blankets, pillow shams, cushion covers, art objects, wall hangers, pot holders, tote bags, costume jewelry, apparels and accessories are some of the never-ending list of items that have been upcycled from denim jeans (Lindzy 2016; Etsy 2016). Upcycling has been carried out in many under-developed and developing countries, but this technique has opened many avenues for young designers, creative artists, manufacturers and retailers, and this concept is spreading all over the globe in its attempt to create products with an image of sustainability.

Fig. 8 Upcycled products from denim jeans (Lindzy 2016; Etsy 2016)

3.4 Constraints in Denim Jean Recycling

Collection and sorting of worn-out jeans is time-consuming and laborious. The material content of the jean is required for sorting, and worn-out labels pose a problem in sorting. A few basic tests can determine the material constituents of jean, e.g., cotton or blends. Currently, many retailers give discounts in new purchases for return of old garments; hence, the problems of acquiring the garments instead of landing in garbage bins are avoided. H & M and other stores are offering 10 % discount voucher on all types of clothing that is returned to the store and also provide 10 % discount coupons for returning empty cosmetic containers of its own brand (Wilbur 2015). Labels, metal parts like rivets and zippers, buttons and leather patches have to be removed manually from the jean before sending it for shredding. Labor is cheap in India, and this process is simple, but in other countries where labor is expensive, this becomes difficult. In most cases, the metal and leather parts are removed and the jean is sent along with the label. Usually, the label contaminates the recycled denim material as it is made of other materials making it difficult during the dyeing process. If any metal parts prevail in the jean which is sent for recycling, it may cause problems to the machinery and process. The heaviness of the metal components in comparison with the denim fiber helps to separate it by the use of gravitation (Science Quest 2016).

Denim jeans are characterized by lapped seams [LSc] as shown in Fig. 9 (Study Blue 2013), which are thick and create problems in shredding and carding. These seams are useful in preserving the quality of the jeans. A small percentage of Lycra is added to create stretch and comfort to the wearer. It is better to separate the Lycra before shredding and cutting, but this can be removed only by chemical recycling. When different colored jeans are recycled together, the resultant yarn is multicolored and can create problems in dyeing. If overdyed uneven color results and may

Fig. 9 Lapped seam LSc
[151]

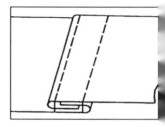

have a bearing on the finished product. Color can be removed by bleaching, but it is not sustainable as effluent is produced; bleaching also reduces the strength and length of the recycled fibers.

3.5 Mechanical Recycling of Denim Jeans

Apparel production involves cutting of fabric into different components which are assembled to form the three-dimensional garment. It has been estimated that about 15–30 % of fabric that is cut for garmenting ends up as cloth scrap and clippings. Conversion of these scraps into insulating materials or nonwoven matting is being done on small levels, and the remaining scraps are sent to landfills. Reports state that £200 million of scrap cotton denim are burned or buried annually. Such reclaimed fibers could be used to produce raw material for developing nonwoven fabrics. Cotton denim is a tough material, tightly woven to make it sturdy for use. Conventional methods of opening denim are very difficult leaving plenty of uno-pened threads, fiber clumps and neps with higher percentage of short fibers, making it difficult to convert into nonwovens by hydroentanglement. Conventional pro-cesses include dry laying or wet laying of fibers followed by hydroentanglement.

The metal parts, labels and leather are removed from the denim jeans and shredded. Mechanical recycling involves the breakdown of denim fabric by thread opening or defibrillation by cutting, shredding, carding or other processes. Cotton, wool and aramid fabrics come under this category, and the fibers obtained are re-engineered into value-added products.

Separating the tufts of fibers into individual fibers and laying them in a parallel form to produce the web are the main objective of carding. Fiber separation is facilitated by mechanical action whereby the fibers are held by one surface and the other surface combs the fibers to form individual fibers. Card clothing, which comprises of needles, wires or fine metal teeth embedded in heavy cloth or metal base, covers a large rotating metal cylinder. The cylinder is partially surrounded by an endless belt with a large number of iron flats. Alternating rollers and stripper rolls cover the top of the cylinder in a roller top card. Figure 10 shows the basic card and the carding action (Voncina 2000).

The licker-in opens the incoming fibers into small tufts and feeds them onto the cylinder. The needles of the two surfaces of the cylinder and flats or rollers are

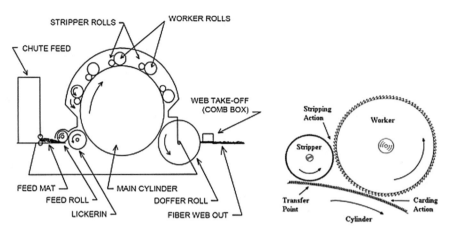

Fig. 10 Basic parts of a card and carding action (Voncina 2000)

inclined in opposite directions and move at different speeds pulling and teasing the fibers apart. The separation of individual fibers occurs between the worker roller and cylinder, while the stripper roller strips the larger tufts and deposits them back on the cylinder. These actions align the fibers into a web below the surface of the needles of the main cylinder. The reprocessed fibers are converted to yarns for woven and knitted fabric manufacture or nonwoven materials for household or technical applications. An re-engineered product for technical applications is designed to meet the requirements and standards set by the different end use industries like construction, automotive, aeronautics and defense.

In many recycling studies a high percentage (50 %) of virgin fibers or non-biodegradable synthetic fibers are being used for manufacturing yarn and fabric which are low in quality and high in cost. When the quantity of virgin fiber additive is high, dyeing has to be performed as the color is diluted. When synthetic fibers are blended, carrier dyeing must be performed to maintain the color increasing costs. Secondly, binders like polyamide-epichlorohydrin, ethylene vinyl acetate EVA and latex are used in percentages ranging from 0.1 to 10 % by weight to hold the short fibers in place. This leads to surface impurities lacking medical sterility and high absorbency restricting its application to industrial wipes, carpet underlay or drying material.

Improved technologies are being used to solve all the problems associated with mechanical denim recycling and develop better quality end products. The cotton denim waste is cut to the required size; the cut material is passed through a sequence of rotary pin cylinders for cutting the material and forming the first web of opened fibers. The cylinders consist of cutting pins fixed to the surface of the cylinder, and the cutting angle may be varied independently to increase/decrease the cutting action depending on the strength and thickness of the fibers. Steam and enzymes are applied to the fibers to remove the surface additives and bacteria.

A fiber balancing process orients the opened fibers into a preferred direction into a web. The opened fibers from the rotary pins or wired cylinders are air lofted and cleaned by a willow cleaner and collected into a condenser to form a lap or batting. The fiber balancing or equalizing process separates the bales of fibers in accordance with size, weight or composition. This enables the manufacturer to pick the required bales of known characteristics to be combined to create the raw material for the desired end product. Fiber equalizing is carried out with the help of a picker or any other means of taking layers from different bales and uniting them sequentially for the required blend. These fibers are sent through a station which applies warm mist of glycerin-based lubricant which softens and untwists the fibers followed by a straightening and combing process. The lubricated and straightened fibers with a preferred uniform directional orientation are sent to the next station for hydroen-tanglement. Cross-lapping of the recycled products gives greater strength and absorbency and could be laminated or coated with a polyurethane laminate barrier making it sterile and suitable for medical applications (Hirsh 2002).

The Fiberization machine at Texperium (NL) has the capacity of working on 200–1500 kg per hour, and the settings of the machine with regard to pins and rollers and compression pressure can be controlled according to requirements. The systems are controlled by air, and filters are used to purify the air. After shredding and opening the material is guided through a machine with finer teeth which removes all bits and pieces and makes the fiber suitable for spinning. Here the fiber is converted to yarns which are lower in quality; the yarns are made stronger by increasing the plies in the recycled yarn, blending with good-quality virgin fibers, by DREF spinning and by the use of better twist.

Denim recycling opens huge opportunities for technical and economic research to arrive at a suitable process. The material obtained by this process is termed as 'recycled cotton' and could be mixed with virgin cotton in different blend ratios as raw material in denim manufacture. Environment conscious customers are attracted toward the 'R- jeans' concept, and manufacturers and retailers are forecasting a growing attractive future for these products (Kobori 2015).

It has been reported that 40–100 % recycled fibers from denim waste is used to make denim fabric suitable for apparel. Figure 11 shows the process flow of obtaining recycled fiber and yarn from denim waste (Ball and Hance 1994). The waste material

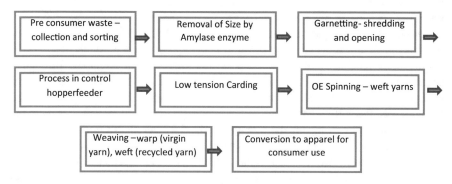

Fig. 11 Process flow for recycled fiber procurement

is chopped into 2″–6″ small pieces. This is followed by enzymatic treatment by adding 1–2 % own of Rapidase XL from International Bio Synthetics for 15 min at 140 °F which converts the starch into sugar which is washed out in the subsequent rinse. The dried scraps are run through high-speed cylinders with steel spikes to break the fabric and bring them into a fibrous condition simulating virgin cotton fiber in bales. This process is known as garnetting. The garnetted material has fiber length ranging from 0.4 to 0.6 in. in length. At this stage lubricants are added to reduce friction among the fibers followed by an opening and cleaning process. The fibers pass through a low tension card, Rieter C4 with conveyor belt, and are converted to a sliver rope and spun into yarns of count 4–16 Ne by open end spinning. Virgin fiber may be added to get better-quality yarn. These yarns are used as weft yarns and woven into fabric which is used for apparel. Dyeing may be undertaken at the yarn, fabric or garment stage to produce a uniform color. The fabric produced has strength and properties that makes it suitable for use by consumer.

3.6 Chemical Recycling in Denim Manufacturing Process

Wash water has long been a problem in any textile processing industry as it contains a high percentage of chemicals that can be recovered and reused. The joint efforts of Dystar and LoopTEC Plant Engineering, a filtration specialist, have brought about recycling of waste water after dyeing, washing and mercerization.

The process of LoopTEC recycling includes cleaning of wash water, dye bath overflow for indigo dyestuff recovery followed by cleaning the polluted indigo dyestuff that is recovered for reuse. The company reports a recovery of 99 % waste water, 99.99 % recovery of indigo dye, 96 % salt recovery and the recovery of indigo concentration can be up to 10 %. The dyeing liquor can be purified to remove sulfur particles and impurities and can be reused after cleaning. This process saves 85 % water which is attained through waste water recovery. The cleaning of caustic soda solution with a pH 14 can be cleaned and recovered (Dystar 2016). Figure 12 shows recovery of indigo dye, wash water and salt in all the three stages of the process, and this leads to a reduction of load and quantity in effluent which makes effluent treatment much easier to achieve zero discharge into the environment.

There are many ways of recycling, but choosing the right techniques to recycle is very important. Many sciences of recycling exist in nature from where inspiration can be drawn for the recycling of products. In today's disposable society it is easier to throw away things, but when we learn to fix things, we learn the art of recycling. Small beginnings in recycling from a large section of people can land up in big benefit to the environment and society. When population increases and the resources are used heavily, even the natural chemical and biological recycling processes become sluggish that we have to start working along with nature to reuse and recycle for turning one thing into another is the magic of recycling.

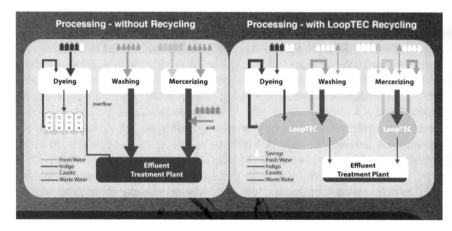

Fig. 12 Role of recycling for recovery and reuse (Dystar 2016)

4 Economics of Denim Recycling

Environmental concerns diminishing landfill space and increasing landfill costs are the concerns of many governments in the global scale. Reducing waste streams at the consumer level is the primary target and may be encouraged by price incentives or recycling. Numerous studies have been conducted on price elasticity and waste disposal at the consumer level, and negative price elasticity has been the highlighted showing that there is an inverse relationship between the price a household pays for waste disposal and the amount of waste disposed from the household. It has been estimated that programs with 80 % recycling rates are viable to produce positive net revenues in terms of sale of recycled materials and cost avoided through recycling. Developing nations like China are on the lookout for recycled materials. Waste material recycled in local conditions is lower in cost than virgin materials and seems to serve the appetite of countries which have a hunger for cheap resources. In the global market, China has become the largest importer of recyclable materials and converts these materials into consumer goods and packaging which is returned as products with recycled labels in countries like Europe and America (The Economist 2007).

The EPA reports that textile recovery from municipal waste contributes to 15.7 % of 14.33 million tons generated in America in 2012 and the rest is discarded (EPA 2014). Despite the fact that denim waste is being converted into many products, about 70 million pounds of scrap denim is being dumped annually in landfills (McCurry and John 1996).

Garments unsuitable for reuse are usually broken down to raw materials for the next use. Companies in Sweden have developed garments from 100 % recycled cotton opening up the market for recycling. Recycled fibers match the quality and price of virgin fibers proving that the next phase of recycling is close loop or circular textiles. However, customer involvement and incentives will encourage the consumers to return their unwanted clothing back for discounts in new purchases

(ACP & I 2016). Another aspect clearly reveals the strong bondage of the consumer to his/her pair of jeans. The real beauty and value of the jeans are discovered after long period of use. A personal relationship is created with appreciation of the person wearing the product. The Mayan weavers of Guatemala are introducing embroidered or woven patterns on to discarded jeans for beautification and restoration, thereby making the product ready for a different buyer rather than the third world under-developed countries. This unique and novel touch will fetch a different outlook for discarded jeans, earn livelihood for many weavers, and also enhances a strong link between the Western world and the Mayans of Guatemala, a place where craftsmanship and connect to clothing are highly appreciated. This paves the way of different end use and market for discarded jeans (Venngoor 2016).

4.1 Efforts for a Circular Economy

The concept of leasing or temporary ownership has been extended by retailers to retain ownership of the raw materials, while consumers keep changing their wardrobe from time to time with moderate cost. Mud Jeans, a Dutch clothing company, has started leasing jeans from 2013 onward, to consumers for a minimum period of one year in return for an annual fee. Once the jeans are returned, the company undertakes the necessary cleaning and repairs to make it ready for another customer.

Clothes that are discarded usually are repaired and sent to under-developed countries for resale as second-hand clothing. This trend is changing slowly, and Patagonia, a leading retailer, is running a second-hand clothing program in four stores in Seattle, Palo Alto, Portland and Chicago where customers can give back their used Patagonia clothes for store credit. A 'worn wear' section in the store sells these goods to new customers.

Retailers like Puma, H & M, American Eagle Outfitters work in collaboration with I:CO, a Swiss reuse and recycling company. Customers who return old garments get discounts for future purchase, and these goods are collected by the reuse company through various collection points. H & M has collected 6 million kg of clothing for such purposes (Vaughn 2014). This effort of donating collected good for reuse, repair or recycling saves resources and helps in the creation of closed-loop production systems. Madewell and 'Blue jeans go green' work hand in hand for recycling worn denim jeans. A vintage truck of denim jeans or two bikes with 40,000 pairs of jeans can provide insulation for forty homes in New York City. These vehicles go around the city collecting old jeans and return them to any one of their 96 outlets. A total of 500–1000 pairs of jeans are required to insulate one home; the customer also gets $20 off for a purchase of a new pair of jeans (Madewell 2015, 2016).

Designing for recycling is the trend, and designing to facilitate repair is the way of extending the life of products in an economy. Certain brands understand their product very well and identify areas which wear and tear easily. The product is well designed that the consumer can easily change the specific part so that they can be easily replaced. A simple example for this is to add extra buttons in denim jackets at the inner side of the button placket to replace them when it is worn out; leather patches are sold in the store while purchasing apparel to add it on in the areas where wear is visible; however, this is very rare as signs of visible wear and tear are part of the worn-out look and style in jeans. Enduring relationships have been built by Levi Strauss by selling a men's brand 'Dockers Wellthread collection' which has special reinforced buttonholes and pockets that is manufactured to last longer and recycled for further use.

4.2 Evaluation of Jeans for Sustainability

The manufacturing of products is dependent on the global supply chain. Any organizations get their products manufactured in places where the cost is low and the working conditions are poor. Certifications from associations like Ethical Trading Initiative and Fair Labor Association have made production systems in factories transparent for scrutiny.

The Ethical Company serves in assisting consumers to pick and choose environmental-friendly products that are based on sustainable manufacture, social responsibility and ethical business. The evaluation of products is based on three general areas—people, animals and environment. While rating jeans for their ethical nature the criteria on which they are adjudged and the ratings secured for different brands are given in Figs. 13 and 14, respectively. The recycling aspect and the end-of-life management are very important features in the environmental reports. The top brands were Calvin Klein, Easy and Falmer with an Ethical index

Fig. 13 Criteria for evaluation—ethical company index

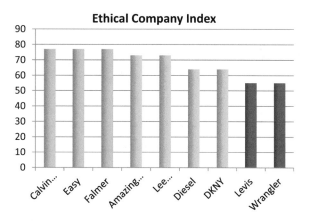

Fig. 14 Jeans ethical rankings (ECO 2014)

score of 77, Amazing Jeans and Lee Cooper with 73, Diesel and DKNY with 64 were considered as middle rated, while Lee, Levi and Wrangler were in the last category with 55 points. In 2005, the Fairtrade Foundation launched certified cotton and products made from fair trade cotton are in high demand; similarly, this foundation will be releasing fair trade jeans for the benefit of the consumers.

4.3 GSTAR—RAW Project

The ocean has six times more plastic than sea life, and the world's biggest landfill is the ocean. Plastics are non-degradable, and they remain in their environment for hundreds of years to become fossils. The plastic from the oceans has woven its way to 'RAW for the Oceans,' a collaborative project which retrieves plastic from the shores of the ocean to transform it into recycled denim (William 2016). About 13,000–15,000 pieces of plastic is dumped into the ocean every day and in a year 6.4 million tons on a global level (Ocean Crusaders 2016). Due to the currents of the oceans the plastics get accumulated at five ocean gyres which contain millions of plastic pieces along which marine life thrives. The first step in the recycling project is to collect plastic pollution from the shorelines of the ocean; this becomes a huge raw material source and millions of tons of virgin plastic are being saved from use along with the reduction of pollution to environment which comes along with the manufacturing process. The retrieved ocean plastic is broken into chips, shredded into fibers and is ready for spinning. The ocean plastic fibers are spun into a strong core yarn with a cotton sheath to form the bionic yarn. GStar supplies this yarn to weaving mills of knitting industries to convert them into RAW for the ocean fabrics. Artistic Milliners, Karachi, are working with bionic yarn provided by GStar for conversion into denim ocean fabrics (Apparel Resources 2015) (Fig. 15).

The economics for plastic bottle recycling is very simple. In 2005, USA recycled 3.3 billion pounds of post-consumer plastics avoiding land fill. Plastic recycling

1	2	3	4	5
Plastic Waste Collection	Plastic chips for shredding	Spinning reclaimed plastic	Weaving or knitting	Denim apparel collection

Fig. 15 Development of recycled denim from ocean plastic waste (Williams 2016)

industry provides jobs to more than 52,000 American workers. Five PET bottles yield enough fiber for one extra-large T-shirt or one square foot of carper or enough fiber fill for a ski jacket. The National Recycling Coalition estimates that if ten million (10,000,000) PET bottles are recycled per week, there is enough recycled fiber to make 104 million (104,000,000) T-shirts per year (National Recycling Coalition 2007).

4.4 Iris Industries—Denimite

Denimite is a new compost made from recycled denim scraps manufactured by NE-based Iris Industries. Denimite is pressed into sheets for further processes like cutting, gluing and sanding or into custom-made molds which may include any products like wallets, drink coasters or rings. The eco-friendly resin is provided by Entropy Resins, and the denim recycled scraps as raw materials are provided by Bonded Logic. The company has posted a video to display the strength of Denimite where a small piece of the product is banged repeatedly against a 161 lb anvil; further, the anvil is also hung from the Denimite product (Baker 2013). New biocomposites will replace less efficient materials, and the company is planning to widen the applications of Denimite to counter tops, tiles, consumer products and automotive applications (Fig. 16).

4.5 Levi Strauss—Jean Manufacturing with Recycled Water

Jean manufacturing has a high rate of pollution in its big water foot print. It has been reported that about 17–20 % of pollution in China is due to the textile industry. In 2014, Levi Strauss & Co. has launched a new method of manufacturing jeans with recycled water in its supplier plant in China. A system of microfiltration was used where the effluent water is passed through a special porous membrane to separate microorganisms and suspended particles from the process liquid (Erdumlu et al. 2012). Levi has stated that 12 million liters of water was saved in a season

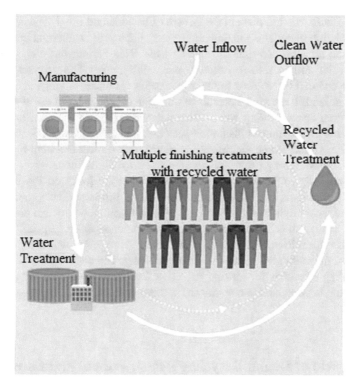

Fig. 16 Levi strauss jean manufacturing with recycled water (Baker 2013)

which is equivalent to fill five Olympic-sized swimming pools (Badore 2014). Based on the recommendations of EPA and WHO, Levi has created a new Water Standard for its suppliers. Finishing is the most important part of jean manufacturing, and water is the key element; Levi has shown its concern to the environment and is planning to take this system to all its supplier industries to save the most precious of all resources water. Apart from these efforts in 2013 Levi used 18,850 pairs of donated jeans to cover the Levi's Stadium and named it 'Field of Jeans' (Yu 2014). When the company remodeled its headquarters in San Francisco, the insulation of the building was prepared with recycled denim Bonded Logic insulation material, thereby extending the life cycle of its products.

4.6 Bonded Logic—UltraTouch Denim Insulation

Post-consumer recycled denim is shredded into scraps and used for the manufacture of insulation material UltraTouch which is Class A fire rated and offers superior thermal and acoustical properties. The recycled fibers are treated with a natural biostat and flash dried to prevent the growth of bacteria and fungi. The construction of the insulation material is the tiny air pockets which provide thermal insulation by trapping the air and also muffles sound waves. This batting has earned multiple LEED credits and is safe to humans and environment. The product is 100 % recyclable and can be returned to the organization for reuse. Savings in energy and reduction in landfill are vital elements of this project. The department of energy has estimated that space heating and cooling consume 44 % of the total energy consumed at home and proper insulation selection can save around 10–30 % of the energy bills (Green Home Guide 2009). Hence, this project can help the environment and society through multiple savings and recycling efforts.

Many cases have been taken up, but all direct our focus on the savings by recycling technologies. A little attention toward all processes and procedures can bring about a new look toward recycling. When education, equity and consumption reduction are combined, the world economy would be more respectful and less harmful to the environment. Innovative research in recycling will bring our attention toward the methods and economics of recycling. The nations all over the world are allocating subsidies for making recycling a profitable business but maybe we should look beyond and move toward a new dose of excitement—zero waste economy.

5   Scenario of Denim Recycling—Past, Present and Future

The concept of sustainability has spread to all walks of life, and consumers today evaluate products on the basis of their features of sustainability. Manufacturing of denim which is very polluting in nature has taken a deep shift toward eco-friendliness that many processes and products have been redesigned to create less harm to the environment and society. Over the past decade India has developed into a powerful sourcing base for denim fabric or product. All major brands like Levis, Gap, Zara, Next, Mango, Calvin Klein are shifting their sourcing base for jeans to Bangladesh. Nearly 75 % of the denim manufactured in India is used for the domestic market, and the remaining is exported to Bangladesh (Apparel Resources 2015). Indian denim industry manufactures a wide range from organic cotton to recycled denim. A study by Technopak Advisors reports that the 300 million pair of jeans is projected to increase to 550–600 million pairs by 2015, and the denim market in India will double to over Rs. 13,000 crores by 2017 (Denim Club 2014). This may be attributed to the young Indian population and their fascination for denim jeans; the boom is also driven by the increasing demand in rural area and small towns and the acceptance of this fabric in workplaces.

Old denim jeans collected under the 'Back to life' program by H & M go to Artistic Milliners in Bangladesh where it is shredded, spun into yarn, woven into fabric; bionic yarn made from reclaimed plastic PET bottles thrown into the ocean is supplied by GStar for developing an exclusive denim product RAW. Vicunha in Brazil uses only BCI cotton for its denim and claims to have attained zero landfill as all waste is reused or recycled. Levi Strauss has introduced a new water recycling standard by which some aspects of garment production use 100 % recycled water to reduce the impact on fresh water resources. Peter England has also introduced various varieties of sustainable non-indigo bottoms which are treated to give a denim look. In China, pollution problems have become a major concern causing drastic reductions in the denim industry as sustainability in process and product is the future of denim.

5.1 Latest Trends Toward Sustainable Denim Jean Manufacturing

Denim is considered to be 'green' when the base raw material is blended with sustainable fibers and the production processes are less water and chemical intensive. According to the Made-by environmental benchmark for fibers, Tencel belongs to Class B along with organic cotton (Made-By 2016), while the SAC Material Sustainability Index (MSI) has given highest rates to Tencel as a cellulosic fiber. A fabric made of 45 % Tencel called Suave is considered to be green and is in production for denim jeans (Lenzing 2011).

Kasav Leather Label, Istanbul, has developed '0 chromium' washable leather with 3–5 % shrinkage, available in above 900 colors. The innovative process makes use of organic tanning agents that does not have any heavy metals in the produced leather as well as the waste water. The natural tanning agents and additives work in a sterilized environment and use lesser time and energy when compared to the conventional tanning methods (Kasiv Leather Label 2016). When converted into a label the edges have a natural vintage effect. The same organization also produces recycled leather by using the discarded material of leather processes and tanneries and converts them to composite materials with adhesive bonding. Shredding and cutting of bits and scraps of leather are followed by combining with resin to form a product similar to leather in feel, weight and durability but environmentally less taxing. Color variations may occur in the finished product as the process is chemical free.

YKK, Atlanta, has eco-friendly processes to showcase the sustainability aspects of its products. Natural materials like salt and stones with a combination of heat have replaced galvanic processes. A unique natural look is found in each single item as there is considerable reduction of chemicals making them environmentally respectful. The conventional anodized aluminum finish for YKK products has been replaced by Anodized Plus Finish which meets the standard AAMA 612 standard for anodized finishes as shown in Fig. 17. Conventional anodized finishes are subjected to hot water or steam sealing process to cover every pore of the anodic layer to prevent staining and overall degradation of the finish. An electrodeposition coating seals the pores completely and protects it from degradation.

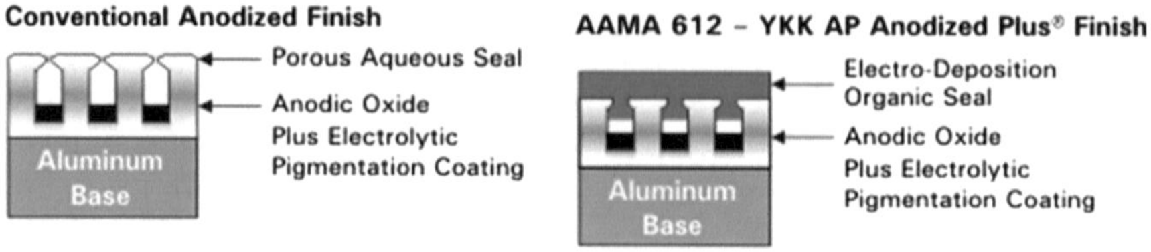

Conventional Anodized Finish

Porous Aqueous Seal

Anodic Oxide

Plus Electrolytic
Pigmentation Coating

AAMA 612 – YKK AP Anodized Plus® Finish

Electro-Deposition
Organic Seal

Anodic Oxide

Plus Electrolytic
Pigmentation Coating

Fig. 17 Sustainable finishes from YKK (2016)

Metalbottoni B 20, a specialist in buttons and accessories, has developed four distinctive styles Monster, No Impact Buttons, Labora and Gummix. A complete ec-sustainable production cycle composed of natural materials and fibers is used for the creation of No Impact Buttons and accessories.

Arvind Limited, India, has made two interesting developments in the field of denim manufacturing based on indigenous legacy. Khadi Denim is a total hand-made product with hand spun yarns, hand hank dyed with natural indigo and woven in a handloom (Rao 2016). The exclusiveness of each garment and the natural imperfections give it a personal and artistic touch. Neo-cord is another product that gives a new dimension to Corduroy. The new Corduroy denims from Arvind come in a wide range of indigo shades, and their unique feature is their fading quality with every wash. This aspect makes them similar to denim jeans and has become popular in the current times.

5.2 Case Studies in Denim Recycling

- **Lee—Denim Recycling**

Lee, one of the world acclaimed premier fashion label, was founded in 1889 in Kansas. The joint efforts of Lee and a nonprofit organization 'A hundred Hands' have given rise to a variety of products from donated jeans showing the care they exhibit for the environment and the under-privileged who create products through recycling craft. Worn-out jeans can be donated at stores in Bangalore, Delhi and Mumbai to provide livelihood to the needy. Further, the e-CREATE range of clothing uses less water than the conventional process and designs jeans from organic cotton and from plastic waste.

Lee uses recycled coffee grounds as a source for fiber, and NeoChaEdge of China has created unique sculptures for the Lee Rethink denim campaign. As part of the sustainability drive, the sculpture that hangs in the Lee Shanghai Flagship store has been created with the raw materials from Lee Rethink process—5000 balls of natural cotton, 200 spools of cotton thread, 50 bags of coffee beans and recycled coffee grounds and 20 articles made from Lees recycling program[196]. Numerous apparel innovations like Zipper Fly jeans, Union-all, Pressed Denim, Loco Jacket and Jelt denim (Apparel Resources 2015) are most striking in the history of Lee, but

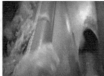

Fig. 18 Mechanical recycling process (Mud Jeans 2016a, b)

the innovation with the focus on sustainability is a major drive to woo customers to their store.

- **Mud Jeans—From Owning to Using Jeans**

Mud Jeans has fabricated a unique scheme of denim recycling where ownership of the jeans is replaced by use and return concept. The consumer's wardrobe will be up to date without much of expenditure, and the company can reuse and recycle the product saving raw material and resources. The promise to its customers has shifted from 'quality assurances' to 'recycling assurances' in their marketing strategy. The circular economy experience, Fig. 18, is shared by the consumers, and they feel honored to be part of the recycling process and a larger share economy. This leaves a lasting impression of their duty toward recycling and brand image building. Mud Jeans is innovating a business model of leasing its organic jeans to ensure safety of raw material supply and to study the possibility of new strategies.

It has been estimated that 30 % of garments have not been worn for a year. Leasing a pair of jeans will prevent this accumulation, and one can be free of guilt to have a desire for newness. Customers become members and rent jeans for a year, and after use three choices are given to keep/swap for new pair/return the jeans in accordance with their choice. The scheme for lease of a pair of Mud Jeans has been shown in Fig. 19. The savings are immense, and the raw material is intact, and the company has a database of loyal customers for formulating new strategies and experiences.

- **REMO—Red Light Denim**

A limited edition of premium denim called Red Light Denim was produced under the brand name REMO with 18 % recycled post-consumer denim collected from the people of Amsterdam (REMO 2014). The REMO database reveals the origin of the product, the past history, the journey of the recycled fibers and the resulting savings. The production history is presented in terms of the steps in production, the name of the manufacturer, date of production, the quantity of recycled fibers as a percentage and the fiber composition as shown in Table 2, and the different stages of production are given in Fig. 20. The savings per kg in terms of carbon dioxide emissions, energy and water are informed to the customer by means of an interactive tag which gives the production history and savings from Red Light Denim/product. The concept of recycling is common, but the transparency in displaying the percentage of recycled content and the savings of the

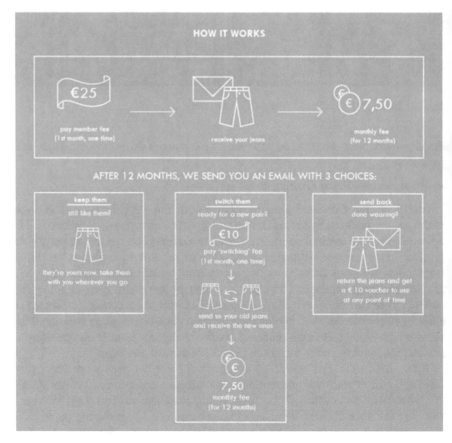

Fig. 19 The scheme for lease of a pair of Mud Jeans (2016a, b)

process is certainly very effective and inspiring. Other products include production of sweatshirts (11 % REM) and Polo Shirts (30 % REM) from used Dutch Military Clothing, Operation Desert Storm; scarves (60 % REM) for Mud Jeans from KLM Air Crew Uniforms; cardigans (49 % REM) for WE Fashion from ABN-AMRO Staff Uniforms; blazers (54 % REM) for UCI officials from discarded PET bottles from Italy.

Denim recycling has been highlighted here by means of a few case studies. The trend today is to take up initiatives to clear all the mess of manufacturing and distribution. In the case of denim manufacturing excessive pollution has been exercised, and it is time that manufacturers look out for new methods to curtail waste instead of finding ways to recycle waste. The waste already created should find ways to avoid incineration and landfills, and this goal can be replaced by thoughts and ideas toward prevention of waste and recycling.

Table 2 Production history of red light denim and savings estimate for sustainable environment (REMO 2015)

Sl. No.	Production step	Produced by	Production data	% of recycled content	Fiber composition ratio recycled cotton: virgin cotton
1.	Collecting	KICI	16.07.2014	REM-100	Post-consumer cotton
2.	Sorting	Wieland	07.08.2014	REM-100	Post-consumer Amsterdam denim
3.	Shredding	ROYO textiles	12.09.2014	REM-100	0: 100
4.	Spinning	ROYO textiles	14.09.2014	REM-18	18: 82
5.	Weaving	ROYO textiles	22.09.2014	REM-18	18: 82

Savings estimate for sustainable environment

Environmental savings	CO_2 (kg)	Energy (kWh)	Water (L)	Reference product
Per kg	0.57	8.20	1255	100 % cotton product

6 Challenges, Prospects in Denim Recycling—Roadmap to Denim Recycling

All industries face immense challenges to withstand the competitive business market. The environmental degradation associated with denim jean production is highly dangerous as the volume of jeans produced and used by consumers today is humongous. Many denim jean manufacturing hubs are in the developing or underdeveloped countries where poverty is high and people look out for jobs to earn their livelihood; the economic crisis compels them to work in such situations where their health and life are at stake. Moreover, manufacturing industries try their level best to withhold the workers paying as little as possible and making them work in miserable conditions. The greatest challenge facing denim recycling is the problem of pollution in all levels of manufacture, and if this can be reduced, it will be a great effort to save the environment.

6.1 Methods to Reduce Environmental Impact of Denim Jeans

• **Design for Recycling**

The DFR concept incorporates recycling and recyclable criteria into the design phase to obtain recycled and recyclable products. Recycled products are made from recycled materials, while recyclable products are manufactured to be recycled after

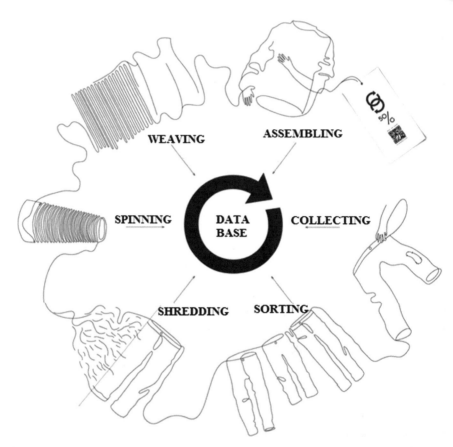

Fig. 20 Stages of production of recycled garments (REMO 2014)

usage or end of life. The most important requirement for recyclable designs is monomaterials, elimination of toxic substances, modular manufacturing for ease of removal or exchange, easily compatible materials, labels or codes can help recycling. The challenges in denim recycling are labels (leather, acetate), metal parts (zippers, rivets and buttons) and thick seams. Buttons can be made detachable if they are designed like cuff links. Care instructions may be printed in pocket inside instead of labels, and seams may be joined by ultrasonic machines or with adhesives. Bleached areas in the garment can represent labels. The current trend is comfort, and the new look is 'athleisure' garments (Lululemon) with high degree of comfort. This becomes a great challenge for jean manufacturers as the garment is not denim and does not have a negative impact on environment.

- **Raw Materials**

From the raw material point of view, cotton can be replaced by other materials which have reduced environmental impact. The key to organic farming is to switch from a chemical-rich farming to a biodiversity-based focus. The techniques include replacement of harmful chemicals, fertilizers and pesticides by compost and rotation of crops for the same land. This way of farming is useful to soil and provides health to farmers. In 2013, ESPRIT partnered with agricultural experts 'Cotton Connect' and Self-employed Women's Association in Gujarat to create the 'Primark Sustainable Cotton Program' and to introduce sustainable farming methods for increased yields and income. About 1251 women farmers were trained, and there was an increase in profits by 211 % which was used for welfare schemes and education of their families (Guang et al. 2013). This program encouraged the organizers to extend this training to another 10,000 female farmers by the next six years.

Organic cotton has been certified by the Global Organic Textile Standard (GOTS) an organization which has set the processing standard for textiles made with organic fibers. This makes the production process traceable and transparent. Organic cotton may cost more, but it ascertains the fact that we care for our planet. The next alternative to organic cotton is 'cotton in conversion.' The general standard for a farmer practicing farming of cotton in a conventional manner to convert to farming of cotton in an organic way is three years. During this time frame, the cotton grown in these lands is known as 'cotton in conversion.' After three years, the cotton from these farmlands will be certified as 'organic cotton' (Walker 2007). Manufacturers who support these farmers will definitely encourage them to continue their farming practices in an organic way. When such support is extended, more cotton will be grown organically leading to healthier soil and healthier farmers.

Organic linen is linen made from the fibers of the flax plant which has been grown without the use of pesticides or fertilizers (Natural Environment 2008). The flax plant is sensitive and susceptible to damage by weeds. Nowadays, linen is made from cotton, hemp and synthetic fibers which resembles linen and termed as 'organic linen.' Crespi 1797, Italy, manufactures organic linen or eco-friendly linen certified that it is made from organic fiber and uses only heavy metal-free dyes (Courtenay 2007). Mikka Works, China, is a fabric and clothing manufacturer who specialize in hemp/organic cotton blends, bamboo, linen and soy fibers (Mikkaworks 2016). EcoLinen of Australia is the first to be certified by Skal Biocontrole International, a control authority that certifies organic products in the Netherlands (Skal 2016). Organic linen fiber is breathable and allows to regulate the body temperature. Rawganique, Canada, works with organic flax, hemp and organic cotton which is chemical free, sweatshop free. Organic linen has worked its way into sustainable clothing and may be considered appropriate for jean manufacturing.

Tencel is a cellulose fiber manufactured from eucalyptus wood pulp. Eucalyptus plants do not need irrigation and grows fast without the use of pesticides. The manufacture of Tencel is a closed-loop process and the solvent used if recovered and reused to certify the process as eco-friendly, water saving and economical.

Recycled raw materials can save water, energy and greenhouse gases. Apparel production gives rise to cotton end bits and cut bits which can become a resource for recycled cotton fibers. If every production center can recycle the end bits and offcuts, then a large savings will result as raw material purchase can be reduced. Similarly, huge savings can be obtained by recycling discarded nylon fabrics and products as virgin nylon production is energy intensive and carbon dioxide emitting in nature. Recycled polyester is obtained from thrown away PET bottles which would otherwise go to the landfill. A second life is given to fibers that are recycled, and this is a great contribution in protecting the environment for the future. A LCA audit has revealed that 75 % water savings in recycled T-shirt production resulted when compared to conventional T-shirt production (Espirit 2016).

- **Manufacturing Alternatives**

The Khadi handmade denim has made its debut in India solely displaying the Indian Khadi cult. Designer Rajesh Prathap Singh and Arvind Mills, India, have collaborated for the birth of the handmade denim Khadi jeans made with handspun yarns and handwoven fabrics. The message 100 % HANDMADE was loud and clear at the Lakme Fashion Week when this collection was displayed. The inspirational theme Khadi has lots of history embedded in it and is an eco-friendly textile that is specially Indian. A fashion show organized by the Denim Manufacturers Association showcased the 'Denim India Made Collection' created by Singh and 11.11/eleven eleven. Apart from these efforts Bangalore-based designer Deepika Govind has launched the 'Denim Green' collection in 2012 after a lot of research and development. Organic cotton is the raw material, handspun yarn dyed with natural indigo with fabric woven on a handloom. This collection is made in classic fits, with tea tree aroma wash or antibacterial wash, and has silver buttons to add a classic touch. The craftsman autographs the handmade denim jeans and adds a code number to add a personal value to the product which carried it into the niche market. With the support of the KVIC, Arvind Mills is earnestly collaborating with many families of craftsmen who are experts in spinning, dyeing and weaving to nurture this concept of Handmade Indian Denim that is a slow fashion which will lay its environmental footprint as an eco-friendly natural product.

Water less technologies are being used to reduce the impact on resources. An average pair of jeans requires 42–45 L of water and may undergo 3–10 wash cycles. Many industries have made changes in their processes like using dry ceramic stones instead of wet pumice stones, combining multiple cycles into one to save water. Under the G2 concept atmospheric air is transformed into plasma which is a blend of active oxygen and ozone molecules. Plasma is used to wash and age the garments. After the air wash, the plasma is converted to purified air and returned into the atmosphere. The G2 air wash machine has a capacity of 50 kg and can wash 3000 pairs of jeans per day. Savings have been estimated to 67 % energy and water, cycle time 55 and 85 % free of chemicals (Vohra 2015).

Another problem in dyeing denim is the use of sulfur dyes. After dyeing around 50 % of the dyes are in the waste water producing water contamination. Sulfur dyes can be recovered and reused which saves money and ends pollution. Wash water from sulfur dyeing can be concentrated through evaporation/filtration; it can be reused by adding chemicals and dyes in small quantities to bring on a standard shade or a lighter shade; sulfur dyes can be titrated with copper sulfate or reducing agents can be added to bring the oxidation reduction potential ORP to a correct level and the alkali can be titrated with 2 end-point titration with HCl and formaldehyde (Mercer 2010). Inexpensive buffers may be used to fix 100 % of the dye which reduces the dye requirement. Cold dyeing methods save energy, eliminate dye waste and improve colorfastness.

Chemical oxygen demand is the byproduct of certain reducing agents used for denim dyeing. Sodium dithionate produces high COD and also causes heavy metal contamination. Hot dyeing methods with non-polluting substances like sugar, dextrins and molasses may be used instead of sodium sulfide reducers, as they break down at high temperature and pH 11 to form hydrogen and alcohol which escape into the air. The normal procedure of chemical oxidation of sulfur blacks is not required; after dyeing with sulfur black it should be allowed to cool in air to allow oxidation, and this should be followed by washing with cold water for further oxidation. The final warm wash will remove the alkalis and residual reducing agents. Selection of the right course of action can save water, dyes and waste water.

Water-saving bleaching technologies are commercially viable, and it can lead to advertisements like low water, low chemical and no chemical finishes. Computer-driven laser technologies provide localized wear, whiskers and patterns without pumice stones. Laser technology is precise and repeatable. However, the equipment is expensive, jeans have to be positioned for treatment, and only one side can be applied at a time, excellent for localized work and not viable for overall bleach. Ozone treatments are the second method of producing a vintage look. Oxygen (O_2) is converted to ozone (O_3). Dampened jeans are exposed to ozone followed by a rinse. Ozone can clean back stains in three seconds and can bleach denim jeans in 15 min which is far lower than chemical and stone washing. Sustainable bleaching is possible as there is savings in energy, chemicals, stones and water. The safety issue with ozone must be taken into consideration as it can cause irritation of eyes, nose and throat (Keshan 2009); this gas corrodes metals, damages plastics and also hardens rubber to cause cracks. Exposure to ozone can cause death, so safety measures are to be adopted. The United States Occupational Safety and Health Administration (OSHA) standards state that the exposure of workers to ozone is 0.1 ppm for 8 h (Bishop 2014). Reports say that ozone treatment reduces water, energy and resources by 50 %.

A technology worth mentioning is the use of ultrasound in manufacturing processes. Natural fibers, made from local plants, can become non-colored by using this technology. The natural fibers are immersed in a solution of sodium permanganate followed by an ultrasound treatment. The manganese oxide molecules settle in the cellulose cavities that naturally occur in the fibers. These molecules react with the dyes and turn them into non-colored particles (Barker 2013). An experiment revealed that dye molecules turned colorless by this technology in a matter of few minutes. This technology can be reversed, and 'structural coloration' can be achieved by placing the nanoparticles directly into the fiber making the dyeing process obsolete. These colors are not affected by UV and hence may not fade.

The concept of structural coloration can be extended to invisibility. A suit embedded with nanoparticles can refract light so that only the background of the person is visible making him invisible to be called as 'sophisticated camouflage.'

6.2 Roadmap to Denim Recycling

The amount of garbage humans produce is enormous and rising day by day. Waste generation is predicted to exceed 11 million tons by 2100 which will be three times more than what it is today. The predicted level of waste would carry serious consequences in the physical and fiscal level. The World Bank report in 2012 warned the people that global waste generation would increase by 70 % by 2025 rising from 3.5 million tons to 6 million tons per day and the global cost for dealing trash will increase from $205 billion a year to 375 billions by 2025 (World Bank 2013).

A thorough sustainability check across the different faces of denim manufacture will help to identify the flaws and prevent them. The focus in denim recycling will be to shift consumer behavior to make recycling clothing a norm, and all phases of the product life cycle are to be taken into account. Collecting old clothing and sending it for recycling only mean that one has not taken part in the recycling process but only helped in facilitating it. If retailers start designing products from old jeans and open a counter in their shop to sell it, then they have started participating in the recycling process in a deeper manner. Further the focus on denim recycling should also go beyond these levels into the secondary level of persons involved in the supply chain.

Suggestions for recycling have been given in design, manufacture, use and end-of-life management shown in Fig. 21. The tips and techniques will enable designers to create designs that are recyclable, manufacturers to take up methodologies for cleaner production and closed-loop economy, savings in the use phase and undertake recycling in the end-of-life management phase. All this can be achieved only if the stake holders work toward a common goal 'zero waste economy.' This goal helps to change the lifestyle of consumers and develop an attitude to take up natural cycles where all the discarded materials will be designed to become resources for the next set of new products. This can be achieved by designing and managing products that avoid the volume and harm created by waste but conserve and recover all resources. Wellthread Pilot is an initiative where designers create 100 % recyclable products that are good in quality and also can be remade into a new product.

The broader sustainability goal for any manufacturer is to create infrastructure and techniques that supports a circular economy. The consumer initiative to return clothes builds their commitment toward environment. The waste management approaches must emphasize on waste prevention rather than addressing the waste in the end. 'Zero waste' shall represent economic alternatives to waste, landfill and incineration and open up a huge opportunity for resource replenishment, business

SUGGESTIONS FOR BETTER RECYCLING – DESIGN TO END OF LIFE MANAGEMENT

DESIGN	MANUFACTURE	USE	END OF LIFE MANAGEMENT
- Mono material - Locally available materials - Materials with low impact on environment - Renewable materials and resources - Preservation of resources - Design for recycling	- Close loop manufacturing - Clean production - Zero discharge and zero landfill - Reuse of parts - Inhouse recycling / member of common recycling center	- Efficient use of energy - Savings in water - Durability - Minimal use of auxillary products - Pollution prevention - Extended use of product by recycling	- Product suitable for recycling - Ease of recycling - Reuse of material - Ease in dismantling - Safe Disposal of toxic substance/ send for recycling

ROLE OF STAKE HOLDERS IN PROMOTING RECYCLING

CONSUMER	RETAILER	GOVERNMENT	GLOBAL
- Selection of goods with recycled content - return of good to store after end of life/charity - reuse if any - try not to send to land fills **MANUFACTURER** - Use of recycled materials - Recycling plan for product after use - Use of resource efficient processes - Inhouse recycling plant or member of common recycling plant - Transparency in supply chain for tracebility in recycling procedures - Submitting end of life management plans and assurances while getting license for a business	- Encourage the customer to buy recycled products - Advertisements and promotional acticities for recycling - Discount for return of product - Showcase the sustainability feature of the retail outlet - Setting brand ambassadors for recycling	- Subsidies for recycled materials - Reduction in taxes for sale of recycled product - Sound policies for recycling - Severe action against polluting industries - Opening centers for recycling - Showcase the level of recycling by statistical information and data - Evaluation and check on impact of recycling - Centers of Research for recycling on state and national scale - Updating and amending laws and regulations as per the current needs in recycling	- International laws to safegaurd nations from unethical manufacturing - Global standards for manufacturing and sourcing in developing and under developed nations - Minimal requirement standards for nations to undertake manufacturing on a global level - International bodies specially to monitor recycling activities - Sharing of research experiences and new techniques to different nations by holding international coferences and platforms for dissemination of knowledge in recycling

ZERO WASTE ECONOMY

Fig. 21 Roadmap to denim recycling

prospects and employment. The greatest resource on hand is the waste that has been generated and accumulated from a long period of time. The first step now is to prevent the generation of waste by adopting the concept of zero waste economy; methods to address the waste accumulated can be taken up for consideration.

6.3   Conclusion

'Denim jeans' which are a classic of all times have spun its way into the lifeline of the people all around the globe. The history of the past still continues to the present and future in authenticating the special magic of blue jeans. Environmental pollution has spiraled to great heights as jean manufacturing is supplying to large sections of the global population. More care for human concerns, especially the work culture, environment, health and safety, is very important for a better future Studies have shown that manufacturing practices in jeans production have caused hazards and the future belongs to sustainable manufacturing and end-of-life management. Literature has proved that existing alternatives like incineration and landfills are also harmful to humans and environment. Designing, manufacturing and using are phases which was carried out for new product development without any other concerns, but today sustainable new product development has become the focus.

We are at a edge with reference to our materials and waste policy and will be stepping of the curb to a new industrial revolution, an economy which just puts forth a simple question—'What if there is not waste?'. This requires a change in attitude and approach as William McDough has stated '*You don't have to filter smokestacks or water. Instead, you put the filter in your head and design the problem out of existence.*' Concepts, philosophies, doctrines and methodologies are available in plenty in the nature around us. All that is necessary is to pick and choose for the right combinations and best results that will safeguard the ecology of our planet. A strong bonding between humans and nature will lead to a better world. Recycling in manufacturing, including denim jeans, and a circular economy with zero waste, is the long-term target set in front of us, and we need to work toward this goal for achievement and prosperity.

> We need a new system of values, a system of the organic unity between humankind and nature and the ethic of global responsibility.
>
> —Mikhail Gorbachev (2004), President, Green Cross International

References

ACP & I. (2016). Going green with blue jeans. http://bluejeansgogreen.org/. Accessed January 20, 2016.

Ahtola, O. T. (1985). Hedonic and utilitarian aspects of consumer behaviour: An attitudinal perspective. *Advances in Consumer Research, 12*, 7–10. http://www.acrwebsite.org/volumes/6348/volumes/v12/NA-12. Accessed April 19, 2016.

Altalhi, T., Kumeria, T., Santos, A., & Losic, D. (2013). Synthesis of well-organised carbon nanotube membranes from non-degradable plastic bags with tuneable molecular transport: Towards nanotechnological recycling. *Carbon, 63*, 423–433. http://www.sciencedirect.com/science/article/pii/S0008622313006246. Accessed April 15, 2016.

Ancona, C., Badaloni, C., Mataloni, F., Bolignano, A., Bucci, A., Cesaroni, G., et al. (2015). Mortality and morbidity in a population exposed to multiple sources of air pollution: A retrospective cohort study using air dispersion models. *Environmental Research, 137*, 467–474. http://www.sciencedirect.com/science/article/pii/S0013935114004071. Accessed April 3, 2016.

Anonymous. (2016a). History of jeans. http://www.historyofjeans.com/jeans-facts/interesting-facts-about-jeans/. Accessed April 19, 2016.

Anonymous. (2016b). Jeans history- origin and invention. http://www.historyofjeans.com/. Accessed April 19, 2016

Apparel Resources. (2015). Expanding application of denim increasing the scope of business. 6 December. http://market.apparelresources.com/sourcing-hub/indian-subcontinent-emerging-as-denim-hub-with-complete-and-compatible-supply-chain/. Accessed April 16, 2016.

Badore, M. (2014). Levi Strauss & Co. launches water-recycling process to make jeans. 26 February. http://www.treehugger.com/corporate-responsibility/levi-strauss-co-launches-water-recycling–process-make-jeans.html. Accessed April 17, 2016.

Baker, B. (2013). Innovative company transforms recycled jeans into coasters, Wallets and rings. 9 December. http://ecowatch.com/wp-content/uploads/2013/12/Untitled.jpg. Accessed April 17, 2016.

Ball, D. L. & Hance M. H. (1994). Process for recycling denim waste – US 5369861. 6 Dec. http://www.google.co.in/patents/US5369861. Accessed April 20, 2016.

Barker, E. (2013). The problem with Indigo. 16 October. http://www.popsci.com/blog-network/techtiles/problem-indigo. Accessed April 20, 2016.

Beck, R. W. (2001). U.S. Recycling economic information study. http://infohouse.p2ric.org/ref/19/18327/fullreireport.pdf. Accessed April 5, 2016

Belly. (2011). Most common finishes for Jeans. January 29. http://www.denimhelp.com/most-common-finishes-for-jeans/. Accessed April 19, 2016.

Berry, M., & Bove, F. (1997). Birth weight reduction associated with residence near a hazardous waste landfill. *105*(8), 856–861. http://www.ncbi.nlm.nih.gov/pubmed/9347901. Accessed April 1, 2016.

Bishop, M. (2014). Ozone finishing for denim reduces environmental impact, processing costs and processing time. 01 August. http://apparel.edgl.com/news/Ozone-Finishing-for-Denim-Reduces-Environmental-Impact,-Processing-Costs-and-Processing-Time94272

Blue Jeans. (2011). The History of Blue Jeans-Impact on America. 31 March. http://bluejeans.umwblogs.org/impact-on-america-2/. Accessed April 20, 2016.

Candy, F. J. (2005). The fabric of society: An investigation of the emotion and sensory experience of wearing denim clothing. *Sociological Research Online*, *10*(1). March 31. Accessed March 26, 2016.

Caufield, K. (2009). Sources of textile waste in Australia. http://www.nacro.org.au/wp-content/uploads/2013/04/TEXTILE-WASTE-PAPER-March-2009-final.pdf. Accessed January 20, 2016.

Comstock, S. C. (2016). The rise and demise of American Blue Jean: Hoe Mexico and East Asia helped make and Unmake twentieth century Icon and National Industry. http://www.ucl.ac.uk/global-denim-project/nn. Accessed March 26, 2016.

Cooper, J. (2016). Happiness: It's not in the jeans. News Release. http://karenpine.com/wp-content/uploads/2012/03/PR-Happiness-its-not-in-the-jeans.pdf. Accessed March 21, 2016.

Courtenay, F. (2007). Crespi 1797: Organic Italian Linen. 13 April. http://www.treehugger.com/sustainable-fashion/crespi-1797-organic-italian-linen.html. Accessed April 20, 2016.

Creek Life. (2016). The death of Denim: How great it would be! https://creeklife.com/blog/environmental-aspects-of-blue-jeans. Accessed April 20, 2016

Cuc, S., & Vidovic, M. (2011). Environmental sustainability through clothing recycling. *Operations and Supply Chain Management*, *4*(2/3), 108–115. Accessed April 5, 2016.

DEEP. (2016). Textiles are so much more than just clothes. http://www.ct.gov/deep/cwp/view.asp?a=2714&q=537718&deepNav_GID=1645. Accessed January 20, 2016.

Denim Club. (2014). Indian denim industry: It's all in the jeans. 10 May. http://www.denimclubindia.org/rsrc/newsPg/news_disp.asp?item_id=4388. Accessed April 16, 2016.

Denim Future. (2016). A complete denim guide for beginner: Washes, finishes & terms. http://www.denimfuture.com/read-journal/a-complete-denim-guide-for-beginner-washes-finishes-and-terms. Accessed April 19, 2016.

Dockery, D. V. V., & Pope, C. A. (1994). Acute respiratory effects of particulate air pollution. *15*, 107–132. http://www.ncbi.nlm.nih.gov/pubmed/8054077. Accessed April 3, 2016.

Dolk, H., Vrijheid, M., Armstrong, B., Abramsky, L., Bianchi, F., Garne, E., et al. (1998). Risk o congenital anomalies near hazardous-waste landfill sites in Europe: The EUROHAZCON study. *Lancet, 352*(9126), 423–427. http://www.ncbi.nlm.nih.gov/pubmed/9708749. Accessec April 1, 2016.

Downey, L. (2014). A short history of Denim. http://www.levistrauss.com/wp-content/uploads 2014/01/A-Short-History-of-Denim2.pdf. Accessed March 16, 2016.

Dystar. (2016). Wash water reuse and indigo recycling on denim dyeing ranges. http://www dystar.com/wpcontent/uploads/2015/11/Indigo_water_resuse_looptec_6_print_Nomarks.pdf. Accessed April 15, 2016.

ECO. (2014). Ethical Jeans. http://www.thegoodshoppingguide.com/ethical-jeans/. Accessed Apri 17, 2016.

Elliot, P., Eaton, N., Shaddick, G., & Carter, R. (2000). Cancer incidence near municipal solic waste incinerators in Great Britain. Part: histopathological and case-note review of primary liver cancer, *82*(5), 1103–1106. Accessed April 3, 2016.

Elliot, P., Hills, M., Beresford, J., Kleinschmidt, I., Jolley, D., Pattenden, S., et al. (1992) Incidence of cancers of the larynx and lung near incinerators of waste solvents and oils in Grea Britain. *339*(8797), 854–858. Accessed April 3, 2016.

Elliot, P., Shaddick, G., Kleinschmidt, I., Jolley, D., Walls, P., Beresford, J., et al. (1996). Cance incidence near municipal solid waste incinerators in Great Britain. *British Journal of Cancer 73*(5), 702–710. http://www.ncbi.nlm.nih.gov/pmc/articles/PMC2074344/. Accessed April 3 2016.

EPA. (2014). Municipal solid waste generation, recycling, and disposal in the United States: Fact and figures for 2012. https://www3.epa.gov/wastes/nonhaz/municipal/pubs/2012_msw_fs.pd Accessed April 13, 2016.

Erdumlu, N., Ozipek, B., Yilmaz, G., & Topatan, Z. (2012). Reuse of effluent water obtained i different textile finishing processes. *Autex Research Journal, 12*(1), 23–26. http://autexrj.com cms/zalaczone_pliki/0005_12.pdf. Accessed April 17, 2016.

Espirit. (2016). Sustainability in practice. http://www.esprit.com/company/sustainability sustainability_in_practice/. Accessed April 20, 2016.

Etsy. (2016). Recycled denim rug. https://www.etsy.com/in-en/market/recycled_denim. Accesse April 15, 2016.

Fischer, C. (2000). Household and municipal waste: Comparability of data in EEA membe countries. http://www.eea.europa.eu/publications/Topic_report_No_32000. Accessed April 2C 2016.

Goldberg, M. S., al-Homsi, N., Goulet, L., & Riberdy, H. (1995). Incidence of cancer amon, persons living near a municipal solid waste landfill site in Montreal, Québec. *50*(6), 416–424 http://www.ncbi.nlm.nih.gov/pubmed/8572719. Accessed April 2, 2016.

Goldberg, M. S., DeWar, R., Desy, M., Riberdy, H. (1999). Risk of developing cancer relative t living near a municipal solid waste landfill site in Montreal, Quebec, Canada. *Archives c Environmental Health, 54*, 291–296. Accessed April 2, 2016.

Goldman, L. R., Paigen, B., Magnant, M. M., & Highland, J. H. (1985). Low birth weigh prematurity and birth defects in children living near the hazardous waste site. *Love Canal, 2(2* 209–223.   https://www.researchgate.net/publication/233783949_Low_Birth_Weight Prematurity_and_Birth_Defects_in_Children_Living_Near_the_Hazardous_Waste_Site_ Love_Canal. Accessed April 1, 2016.

Gorbachev, M. (2004). The road to a sustainable environment and a safer world: A call for globa glasnost. http://www.socsci.uci.edu/files/announcements/cpb/gorbachev.htm. Accessed 23 March

Gordon, S., & Hsieh, Y. L. (2007). *Cotton: Science and technology*. England: Woodhea Publishing Limited. Accessed January 20, 2016.

Green Home Guide. (2009). Choosing the best insulation delivers energy savings. http://www greenhomeguide.com/know-how/article/choosing-the-best-insulation-delivers-energy-savings. Accessed April 17, 2016.

Guang, L., Mingzhuo, J., & Guang, L. (2013). The denim capital of the world: So polluted you can't give the houses away. 13 August. http://www.primark.com/en/our-ethics/environment/raw-materials. Accessed April 20, 2016.

Gustavsson, P. (1989). Mortality among workers at a municipal waste incinerator. *American Journal of Industrial Medicine, 15*(3), 245–253. http://www.ncbi.nlm.nih.gov/pubmed/2929614. Accessed April 3, 2016.

Hasan, K. (2013). Environmental damage in Bangladesh. 14 July. http://www.nytimes.com/slideshow/2013/07/14/world/asia/07142013BANGLADESH-9.html. Accessed April 20, 2016.

Hegarty, S. (2012). How jeans conquered the World. BBC World Service. http://www.bbc.com/news/magazine-17101768. Accessed March 15, 2016.

Hirsh, G. F. (2002). System and method for reconstituting fibers from recyclable waste material—US 6378179 B1. http://www.google.co.in/patents/US6378179. Accessed April 6, 2016.

Kasiv Leather Label. (2016). http://www.kasivetiket.com/tr/8-sifir-krom–oeko—tex-leather/. Accessed April 16, 2016.

Katsouyanni, K., Touloumi, G., Spix, C., Schwartz, J., Balducci, F., Medina, S., et al. (1997). Short-term effects of ambient sulphur dioxide and particulate matter on mortality in 12 European cities: Results from time series data from the APHEA project. *Air Pollution and Health: A European Approach, 314*(7095), 1658–1663. http://www.ncbi.nlm.nih.gov/pubmed/9180068. Accessed April 3, 2016.

Keshan, S. P. (2009). Ozone: A tool for denim processing. http://www.fibre2fashion.com/industry-article/4146/ozone-a-tool-for-denim-processing?page=1. Accessed April 20, 2016.

KNITcMA. (2011). Vision (Agenda). http://www.knitcma.com/KNITcMA_HtmX/Vision.htm. Accessed March 29, 2016.

Kobayashi, V. (2013). Raw materials for blue jeans. 3 March. http://www.designlife-cycle.com/denim/. Accessed April 19, 2016.

Kobori, M. (2015). Levi Strauss & Co: Bring us your old jeans. http://ecowatch.com/2015/07/22/levis-bring-us-your-old-jeans/. Accessed April 7, 2016.

Kriger, C. E., & Connah, G. (2006). Cloth in West African history. UK: Altamira Press, ISBN 0-7591-0422-0. Accessed April 20, 2016.

Lenzing. (2011). Tejidos Royo: New concepts. 21 February.http://blog.lenzing.com/2011/02/tejidos-royo-new-concepts/. Accessed April 16, 2016.

Levi Strauss & Co. (2015). The life cycle of a jean. January 24. http://levistrauss.com/wp-content/uploads/2015/03/Full-LCA-Results-Deck-FINAL.pdf. Accessed April 19, 2016.

Levi's. (2016). What's your curve ID. http://www.levi.com/GB/en_GB/women/fit-guides/curve-id-tool. Accessed April 19, 2016.

Lindzy, V. (2016). Denim upcycled—rag rug—loom woven. https://www.etsy.com/in-en/listing/243499935/denim-upcycled-rag-rug-loom-woven? Accessed April 15, 2016.

Little, D. (2007). *Denim: An American Story*. USA: Schiffer Publishing Ltd. 1 July. http://www.amazon.com/Denim-American-Story-Schiffer-Book/dp/0764326864. Accessed March 26, 2016.

Made-By. (2016). About the benchmark for fibers. http://www.made-by.org/consultancy/tools/environmental/. Accessed April 15, 2016.

Madewell. (2015). Our national denim recycling drive, by the numbers. 8 October. https://blog.madewell.com/2015/10/08/our-national-denim-recycling-drive-by-the-numbers/#more-8974. Accessed April 15, 2016.

Madewell. (2016). Recycle you old jeans with us. https://www.madewell.com/madewell_feature/DENIMDONATION_sm.jsp. Accessed April 15, 2016.

McCurry & John W. (1996). Blue jean remnants keep homes warm. *Textile World, 146*(10), 84–89. http://connection.ebscohost.com/c/articles/9702161537/blue-jean-remnants-keep-homes-warm. Accessed April 15, 2016.

Mercer, H. (2010). Denim pollution—solutions to sulphur dyeing wastes. 15 May. http://www.denimsandjeans.com/denim/manufacturing-process/denim-pollution-solutions-to-sulphur-dyeing-wastes/. Accessed April 20, 2016.

Mikkaworks. (2016). Fabrics and clothing production. http://www.mikkaworks.com/. Accessed April 20, 2016.

Mondal, M., Gohs, U., Wagenknecht, U., & Heinrich, G. (2013). Polypropylene/natural rubber thermoplastic vulcanizates by eco-friendly and sustainable electron induced reactive processing. *Radiation Physics and Chemistry, 88*(0), 74–81. http://adsabs.harvard.edu/abs/2013RaPC...88...74M. Accessed April 15, 2016.

Mud Jeans. (2016a). Lease a jeans. http://www.mudjeans.eu/lease-a-jeans/. Accessed April 15 2016.

Mud Jeans. (2016b). The Recycle Process. http://www.mudjeans.eu/recycle-tour/the-recycle-process/. Accessed April 15, 2016.

National Recycling Coalition. (2007). Conversionator—recycling calculator. http://philmang.com/work/nrc/shell_pepsi.html. Accessed April 17, 2016.

Natural Environment. (2008). What is organic linen. 24 January. http://www.natural-environment.com/blog/2008/01/24/what-is-organic-linen/. Accessed April 20, 2016.

Ocean Crusaders. (2016). Plastics ain't so fantastic. http://oceancrusaders.org/plastic-crusades/plastic-statistics/. Accessed April 17, 2016.

Op, G. (2014). Life cycle of a jean and our clothing footprint. https://globalaspect.wordpress.com/2014/01/24/life-cycle-of-a-jean-and-our-clothing-footprint/. Accessed April 19, 2016.

Paul, R. (2015). Denim—manufacture, finishing and applications. UK: Woodhead Publishing. ISBN 9780857098436. Accessed April 5, 2016.

Peck, J., & Wiggins, J. (2006). It just feels good: Customers' active response to touch and its influence on Persuasion. *Journal of Marketing, 70*, 56–69. http://www.scodix.com/wp-content/uploads/2012/12/It-Just-Feels-Good-Customers-Affective-Response-to-Touch-and-Its-Influence-on-Persuasion.pdf. Accessed April 19, 2016.

Rahman, O. (2011). Understanding consumers' perceptions and behaviour: Implications for denim jeans design. *Journal of Textile and Apparel, Technology and Management, 7*(1), 1–16. http://ojs.cnr.ncsu.edu/index.php/JTATM/article/viewFile/845/909. Accessed April 19, 2016.

Rao, S. I. (2016). Potentially revolutionary. http://www.borderandfall.com/karigar/khadi-denim-revolutionary-textile/. Accessed April 16, 2016.

Rapiti, E., Sperati, A., Fano, V., Dell'Orco, V., & Forastiere, F. (1997). Mortality amongst workers at municipal waste incinerators in Rome: A retrospective cohort study. *American Journal of Industrial Medicine, 31*, 659–661. Accessed April 2, 2016.

REMO. (2014). It's not over for used garments, in fact it's just the beginning. http://www.joinremo.com/. Accessed April 19, 2016.

REMO. (2015). Red light denim, carrying the free spirit of Amsterdam inside. http://www.joinremo.com/case-studies/. Accessed April 18, 2016.

Robinson, T. E. (2003). Clothing behaviour, body cathexis, and appearance management of women enrolled in a commercial weight loss program. Dissertation. https://theses.lib.vt.edu/theses/available/etd-08012003-155510/unrestricted/TR-ETD.pdf. Accessed April 19, 2016.

Rushton, L. (2003). Health hazards and waste management. *British Medical Bulletin, 68*, 183–197. http://bmb.oxfordjournals.org/content/68/1/183.full. Accessed April 9, 2016.

Science Quest. (2016). Separating mixtures. http://www.wiley.com/legacy/Australia/PageProofs/SQ7_AC_VIC/c05SeparatingMixtures_WEB.pdf. Accessed April 9, 2016.

Schmitt, B. H., & Simon, A. (1997). *Marketing aesthetics: The strategic management of brands, identity and image*. New York: Free Press. Accessed April 19, 2016.

Schorlemmer, C. (1874). *A manual of the chemistry of the carbon compounds or organic chemistry*. London: Macmillan and Co. Accessed April 20, 2016.

Shore, M. (1994). The impact of recycling on jobs in North Carolina, University of North Carolina. http://infohouse.p2ric.org/ref/24/23720.pdf. Accessed April 5, 2016.

Skal. (2016). EC legislation. https://www.skal.nl/home-en-gb/about-skal/ec-legislation/. Accessed April 20, 2016.

Steingruber, E. (2004). *Indigo and indigo colorants, Ullmann's encyclopedia of industrial chemistry*. Weinheim: Wiley-VCH, doi:10.1002/14356007.a14_149.pub2. Accessed April 20, 2016.

Study Blue. (2013). Seam classification. https://www.studyblue.com/notes/note/n/amad-231-study-guide-2013-14-baytar-/deck/8693137. Accessed March 15, 2016.

Sullivan, J. (2006). *Jeans: A cultural history of an American Icon*. New York: Gotham Books. Accessed March 15, 2016.

The Economist. (2007). The truth about recycling. 7 June. http://www.economist.com/node/9249262. Accessed December 12, 2016.

Trash to Trend. (2016). Trash to trend—Concept. http://trash-to-trend.myshopify.com/pages/concept. Accessed March 29, 2016.

Upadyayay, D., & Ambavale, R. (2013). A study on preference with reference to denim jeans in female segment in Ahmedabad City. *International Journal of Management and Social Sciences (IJMSSR)*, 2(4), 153–159. Accessed April 19, 2016.

Vaughn, J. (2014). Five new ways the circular economy can build brand experience. http://www.theguardian.com/sustainable-business/five-ways-circular-economy-brand-experience. Accessed April 15, 2016.

Venngoor, M. (2016). Denim economics. http://www.markvennegoor.nl/current-projects/denim-economics/. Accessed April 15, 2016.

Vianna, N. J., & Polan, A. K. (1984). Incidence of low birth weight among Love Canal residents. *Science*, 226(4679), 1217–1219. Accessed April 1, 2016.

Vohra, A. (2015). The green denim. 19 April. http://www.financialexpress.com/article/industry/companies/the-green-denim/64861/. Accessed April 20, 2016.

Voncina, B. (2000). Recycling of textile materials. http://www.2bfuntex.eu/sites/default/files/materials/Recycling%20of%20textile%20materials_Bojana%20Voncina.pdf. Accessed April 7, 2016.

Vrijheid, M., Dolk, H., Armstrong, B., Abramsky, L., Bianchi, F., Fazarinc, I., et al. (2002). Chromosomal congenital anomalies and residence near hazardous waste landfill sites. *Lancet*, 359(9303), 320–322. http://www.ncbi.nlm.nih.gov/pubmed/11830202. Accessed April 1, 2016.

Walker, A. (2007). 100 % organic cotton in conversion. 15 August. http://www.indigoclothing.com/blog/100-organic-cotton-in-conversion/. Accessed April 20, 2016.

Weber, C. (2006). Me and my Calvins. The New York Times. August 20. http://www.nytimes.com/2006/08/20/books/review/20Weber.html?_r=0. Accessed April 20, 2016.

WHO. (2000). Methods of assessing risk to health from exposure to hazards released from waste landfills. 12 April. http://apps.who.int/iris/bitstream/10665/108362/1/E71393.pdf. Accessed April 1, 2016.

Wilbur, H. (2015). Rags for riches: Clothing brands are offering discounts for used clothing. http://mashable.com/2015/11/08/recycle-used-clothing/#5ND0c5aDIGqs. Accessed April 16, 2016.

William, P. (2016). Project weaving the way to cleaner oceans. http://rawfortheoceans.g-tar.com/#!/tagged/project/0. Accessed April 17, 2016.

Williams, P. (2016). Raw for the oceans is a purpose made into a clothing line. http://rawfortheoceans.g-star.com/#!/post/112524429531. Accessed April 15, 2016.

World Bank. (2013). Global waste on pace to triple by 2100. 30 October.http://www.worldbank.org/en/news/feature/2013/10/30/global-waste-on-pace-to-triple. Accessed April 21, 2016.

Wu, J., & Delong, M. (2006). Chinese perceptions of western-branded denim jeans: A Shanghai case study. *Journal of Fashion Marketing and Management*, 10(2), 238–250. http://www.emeraldinsight.com/doi/abs/10.1108/13612020610667531. Accessed April 19, 2016.

WWF. (2008). The 2008 living planet report. http://wwf.panda.org/about_our_earth/all_publications/living_planet_report/living_planet_report_timeline/lpr_2008/. Accessed April 5, 2016.

YKK. (2016). Anodized plus aluminium finishes. https://www.ykkap.com/commercial/performance-product-lines/anodized-plus-aluminum-finishes/. Accessed April 15, 2016.

Yu, N. (2014). Research levi's. http://yuhiunamtextileinnovation.blogspot.in/2014/11/research-levis.html. Accessed April 17, 2016.

Printed in the United States
By Bookmasters